Birkhäuser

# Modern Birkhäuser Classics

Many of the original research and survey monographs in pure and applied mathematics published by Birkhäuser in recent decades have been groundbreaking and have come to be regarded as foundational to the subject. Through the MBC Series, a select number of these modern classics, entirely uncorrected, are being re-released in paperback (and as eBooks) to ensure that these treasures remain accessible to new generations of students, scholars, and researchers.

Gennady E. Gorelik
Victor Ya. Frenkel

# Matvei Petrovich Bronstein

## and Soviet Theoretical Physics in the Thirties

Translated by Valentina M. Levina

Reprint of the 1994 Edition

Gennady E. Gorelik
22 Chestnut Pl., Apt 406
Brookline, MA 02445
USA
gorelik@bu.edu

Victor Ya. Frenkel†
A.E. Ioffe Physicotechnical Institute
St. Petersburg, Russia

2010 Mathematics Subject Classification: 01A70, 81-03, 83Fxx

ISBN 978-3-0348-0199-7 e-ISBN 978-3-0348-0200-0
DOI 10.1007/978-3-0348-0200-0

Library of Congress Control Number: 2011937696


Originally published under the title Matvei Petrovich Bronshtein, 1906-1938, in the series "Nauchno-biograficheskaya literature" by Izdatel'stvo "Nauka", Moskva 1990.
Translated from the Russian by Valentina M. Levina and published as volume 12 under the same title in the Science Networks. Historical Studies series by Birkhäuser Verlag, Switzerland, ISBN 978-3-7643-2752-1
Reprint 2011 by Springer Basel AG



Printed on acid-free paper

Springer Basel AG is part of Springer Science+Business Media

www.birkhauser-science.com

# Contents

Matvei Petrovich Bronstein, 1906–1938

# Foreword

The true history of physics can only be read in the life stories of those who made its progress possible.

Matvei Bronstein was one of those for whom the vast territory of theoretical physics was as familiar as his own home: he worked in cosmology, nuclear physics, gravitation, semiconductors, atmospheric physics, quantum electrodynamics, astrophysics and the relativistic quantum theory. Everyone who knew him was struck by his wide knowledge, far beyond the limits of his trade. This partly explains why his life was closely intertwined with the social, historical and scientific context of his time.

One might doubt that during his short life Bronstein could have made truly weighty contributions to science and have become, in a sense, a symbol of his time. Unlike mathematicians and poets, physicists reach the peak of their careers after the age of thirty. His thirty years of life, however, proved enough to secure him a place in the *Greater Soviet Encyclopedia*. In 1967, in describing the first generation of physicists educated after the 1917 revolution, Igor Tamm referred to Bronstein as "an exceptionally brilliant and promising" theoretician [268].

A collection of major works on the theory of gravitation, put out in 1979 to mark Einstein's birth centenary [90], contained an article by Bronstein. This was a pioneering work on quantizing gravitation that produced important physical results, the details of which we shall discuss later. The article marked Bronstein's peak, the significance of which goes beyond the history of physics. Today the quantum theory of gravitation occupies a special place in fundamental physics; indeed, the last fifty years have altered its interpretation and charted different ways to its solution, yet one of the characteristics discovered and described by Bronstein in 1935 is still beyond the scope of contemporary theoreticians. We are thinking of the fact that the classical relativistic theory of gravitation (the General Relativity theory) is incompatible with the quantum theory that Bronstein proved by physical analysis. Bronstein was the first to realize that the true synthesis of relativity and quantum ideas (the quantum theory of gravitation included) would require a major reconstruction of the concepts of time and space.

It is common knowledge that by altering these fundamental concepts, Einstein produced outstanding results in physics (he connected space and time in the theory of relativity and curved space-time in the theory of gravitation). No wonder, therefore, that those who came after him were tempted on many occasions to forecast "space-time" solutions to fundamental physical problems. Each time their hopes were buried under new physical ideas and experiments. Bronstein's 1935 quantum-gravitational forecast has withstood the test of time. One may argue that forecasts are not, strictly speaking, an integral part of a physical theory; today it has become clear that the quantum theory of gravitation is destined to be a major achievement of theoretical physics and promote our ideas about space and time.

Illustrious lives are more than gemstones incrusted in mankind's history – they provide us with an insight into it. It was Bronstein's fate to live in one of the most tragic and fascinating periods in Russian history. Much of his life was shaped by the time he lived in. It seems that the biographical genre is guided by a principle akin to the quantum uncertainty principle formulated in the thirties: any attempt to isolate the principal figure from his historical milieu deprives his life of purpose and meaning. As quantum mechanics employs Planck's universal constant so "biographical mechanics" operates with its hero's "socializing constant", that is, a number of people, both friends and foes, who were close to him.

We have interviewed some twenty people from Byurakan in Armenia, Minsk in Byelorussia, Sverdlovsk in Russia and Oxford in Great Britain. It was a pleasant surprise that prominent scientists and highly placed administrators responded promptly to our requests to share their reminiscences about Matvei Bronstein. His former colleagues and friends from Moscow and Leningrad seemed to remain under the spell of his vivid personality. It was more than a natural desire of not-so-young people to recreate the past – they were obviously happy to speak about this remarkable man and fine scientist.

This book is based on stories from those who remembered Bronstein as "Mitya" and, from his student years, as "The Abbot". There were people who referred to him by his initials M.P., and by his name and patronymic, Matvei Petrovich (according to the Russian custom). These people still keep in their personal libraries his works and books with his personal dedications, summaries of his lectures and his poetic improvisations jotted down in his own hand.

Bronstein's lectures and articles, their memory still alive among his contemporaries betrayed a talented writer and profound scientist. It seems that he was fascinated with thoughts and ideas clothed in written and spoken words; no wonder teaching and popular science attracted him so much. His first scientific work appeared when he was 18 and not yet a student; his first popular scientific effort was published four years later.

He was a great lecturer and master of scientific explanation thanks to the profound knowledge and enthusiasm of a teacher and the talent of a writer. This urged him to write popular science books for children and young people, addressing the widest and most responsive group of readers. His books, which are as much fine literature as they are brilliantly presented science, are still very popular and have run into many editions. Time has not yet shelved them among other curiosities of literary and scientific history.

Bronstein's scientific articles bear the stamp of his literary talent. Unlike many shallow publications, where stylistic means are employed to camouflage a poverty of ideas, his elegance of expression was not an aim but a means of channelling the reader's attention.

Today scientific publications are written in a boring and dull style, as if authors are afraid of treading beyond the narrow path between "if ... then" and mathematical formulae and thus unwittingly demonstrating that science is a boring and impersonal trade. True, some of this can be accounted for by the natural desire to be precise and unambiguous. Any physicist promptly grows accustomed to this style. Bronstein refused to obey these rules, or, rather, he ignored them. As can be expected of theoretical physics in the twentieth century formulae took much of the space in his articles. At the same time, he quoted Newton in his article on the transmutation of photons into gravitons, employed a German proverb to summarize a major point, replaced the cautious "there are grounds to believe" with the positive "I believe", etc. Those who knew Bronstein well were not put off or shocked by his manner of writing: "He was M.P. and he knew how to present his ideas." This seemed to be part of his many-sided and harmonious personality.

Both prominent physicists who have already earned themselves a place in the history of physics and people far removed from the world of science were aware of Bronstein's talent for human communication. We were greatly encouraged by their reminiscences. Though not all names are mentioned in the text (if the author is obvious from the context), to avoid unnecessary complications we have deemed it necessary to say that we are greatly indebted to M. Bronstein's brother Isidor, his widow, Lydia Chukovskaya, his fellow students V. Ambartsumyan and E. Peierls (Kanegisser), his friends G. Egudin and S. Reiser, A. Migdal who did post-graduate course under him and Ya. Smorodinsky who was his student.

We are deeply grateful to all who shared their memories of Matvei Bronstein and the developments in physics in the twenties and thirties. They were A. Anselm, M. Veselov, S. Vonsovsky, I. Gurevich, L. Gurevich, D. Ivanenko, I. Kikoin, A. Kozyrev, M. Korets, R. Peierls, L. Pyatigorsky, I. Rozhansky and V. Saveliev.

We thank A. Migdal for his critical comments even if we sometimes disagreed with them, and also B. Bolotovsky, V. Vizgin, A. Kozhevnikov, A. Livanova, B. Yavelov and A. Yankovsky. Our special gratitude goes to G. Savina, who found documents indispensable for our book.

A complete list of Bronstein's works can be found at the end of the book. There are no references for his works if the context is clear enough. All contributions in square brackets belong to the authors.

Sections 2.4, 3.4, 3.9, 5.1, 6.3 were written jointly; Sections 2.1, 3.1, 3.2, 3.3 and 3.10 were contributed by V. Frenkel, the rest by G. Gorelik.

# Chapter 1
# Childhood and Youth. Road to Science

Matvei Bronstein was born into a typical family of Russian-Jewish middle intelligentsia in the Ukraine. His ancestors lived within the Pale. His father, Petr, was a doctor. Born into a family of a petty trader he, unlike many other Jewish boys, managed to finish a course in a classical school in a small Ukrainian town and the Department of Medicine of Kiev University. His mother, Fanny, could barely read and write. She was a kind and considerate woman engrossed in the life of her husband and three children – twin sons Matvei and Isidor and a daughter Mikhalina, four years older than her brothers. The boys were born on December 2, 1906, in Vinnitsa, a regional center in central Ukraine. There they spent the first nine years of their lives.

It is known that twins stand less of a chance of being lavishly endowed by nature and can count on less spectacular achievements. In the case of Bronstein twins this proved to be true only of the boys' physical constitution; their intellectual power was far above the average. Although they were not identical twins, it did not prevent them from being very close, especially in childhood.

Not being very much interested in science and literature their parents nevertheless strongly believed in education and reading. Mikhalina attended a grammar school; the younger boys (unusually quiet and inquisitive) were given a lot of books to read. They remembered *deluxe* editions of books on astronomy, history, travel stories, etc.

The family was indifferent to religion, partly because of the father's medical education and the general mood of the time. He held no strong atheistic convictions either, so the Jewish religious rites and customs were quietly ignored. The very names of the children showed that the family was distancing itself from its ethnic background – Matvei was called at home by the Russian name Mitya. His closest friends and family knew him by this name all his life.

In August 1914 this well-ordered (and, as the boys thought, "dull") life was disrupted. As a doctor their father was promptly drafted, and the family found it difficult to make ends meet. Profiting from the fact that her husband's higher education permitted his family, unlike simple Jews, to live outside the Pale, Fanny, together with her children, moved to Kiev to seek support from her father, a manager in a rich merchant firm.

A religious man, the grandfather found the neglect of his grandsons' religious education scandalous. His efforts to enlighten them posed the first philosophical dilemma for the brothers. Resolving this dilemma was an important stage in their intellectual and spiritual maturing. To verify the truth of the religious postulates supplied by the grandfather they staged an experiment: they stood in the middle of a room and loudly declared that God was a fool. Since no immediate retribution followed, they concluded that there was no God. Any theory needs experimental substantiation, otherwise thirteen-year-olds would find it hard to be convinced. What was more important, however, was the intellectual quest for truth that was imprinted on their minds. The grandfather's authority was shaken, and he left the boys in peace. This was all the more easy since they never disturbed their elders with misbehavior: all their time was taken up by reading.

The boys never attended a secondary school in their home town of Vinnitsa – they were too small to be admitted. In Kiev there were two other reasons: the percentage of Jewish children admitted into grammar schools in Kiev was even lower than in Vinnitsa; and the family did not have enough money to educate two boys. These circumstances could be overcome singly: combined they were a solid obstacle. Their sister was attending the best grammar school in Kiev – she had been admitted because her father was in the army-in-the-field. Young as they were the boys' intellectual potential was enough for systematic studies. Their mother and grandfather decided that they could study at home and sit for exams at a grammar school. A French teacher was hired, other subjects were studied independently in textbooks. The boys passed exams with honors for the first three years yet the studies left few, if any, impressions on their minds: the books that they read opened wider vistas.

In 1917 the revolutionary storm destroyed the old social order; grammar schools and high schools gave way to the unified school. Soviet power came to Kiev in the summer of 1920, after German, Polish and White Guard occupations, so the changes to the educational system came late as well.

The Bronstein brothers were not destined to study in the new school either; by that time they had learned how to educate themselves and to select the facts and knowledge they needed.

They were much more impressed by what they read in books than by what was going on around them. While living in downtown Kiev they had no chance to observe nature. When the family, driven by hunger, moved to the countryside to live off the land, the boys exhibited no interest in nature. After a while, they had to move back to Kiev to be secure in the turbulent times of the Civil War.

They were too young and too protected by the family to realize that they were witnessing historic events. They remained engrossed in their books, especially in the history of mankind, efforts to probe the secrets of nature and to change the world. These ideas stimulated their minds more than simple contemplation of the sky and trees; it was much more exciting to learn how Nature was functioning. The boys had already learned that nature was composed of stars, planets, crystals, atoms and electricity.

An interest in everyday natural phenomena came much later, when Mitya's thirst for knowledge had been somewhat quenched. He learned to tell a pine from a fir and wheat from oats, to ride a bicycle and to row. This proved to be fun.

In the early twenties, however, the boys were consumed by the desire to learn and understand; there was small room for fun in their minds and souls.

At that time they were indiscriminate readers – the history of Egypt was as fascinating as the theory of sets. A shallow mind could have become superficial, but Mitya, with his powerful intellect, brilliant memory and self-discipline, acquired wide and profound knowledge which amazed friends in his adult years.

The brothers' interest in books never abated, although with time they chose different subjects. Mitya developed an interest in physics and astronomy.

The books they read varied greatly. Those that appeared before the revolution seemed to have been published in the past century: good paper, gilded covers and pretty pictures. In one of them an elegant young man with long curly hair was sitting in an apple orchard. The Moon was in the background and an apple was in the foreground. The young man was Newton trying to guess why an apple fell to the earth while the moon remained in the sky. The author explained in a popular and elegant French style how Newton had discovered the law of universal gravity, how he had realized that but for its huge velocity the Moon would not have tumbled down to the Earth. The French author calculated that this tumble would have taken four days, 19 hours, 54 minutes and 57 seconds. "It is for the reader to imagine all the horrors of this", he concluded.

There were other books printed on inferior paper that had no pictures and even no hard covers. They were published in Moscow, Petrograd, Odessa and even Berlin (this was the time when books for Soviet Russia were printed in Germany). They told a different story about universal gravitation and used a different language.

They recounted stories of two-dimensional creatures travelling over a sphere; they mentioned the curvature of space and time and explained that according to Einstein's theory a man gazing forward into space could see his own ears. Some books were overenthusiastic about the revolutionary physical theories that changed the time-honored ideas about space, gravitation, atoms and light and could be likened to the revolutionary social changes.

The home library could no longer satisfy the wide-flung scientific interests of the twins. It was practically impossible to buy scientific books, besides, the family no longer had money to waste on books. The boys had to turn to lending and reference libraries; the city central library was their favorite haunt until it closed its doors to schoolchildren. Mitya had to finish Flammarion's *Astronomy for Everyone* (quoted above) in the library of the Academy of Sciences, which had no age qualifications for its readers.

When they turned sixteen it became obvious that the twins had to learn some useful trade. In 1923 they entered a specialized secondary school to study electrical engineering but had to abandon their studies to help the family. Their father proved unable to work outside a state medical establishment that paid little money. The boys started working at a plant.

At seventeen, young people are very inquisitive and are looking for their own roads in life. Little by little it had become obvious that nature had not divided the genetic inheritance equally between the twins. Matvei was much better in the exact sciences: it was quite by chance that his brother discovered that Matvei had mastered the trigonometry that he himself never studied. This seemed strange because their starting conditions had been completely identical.

In 1924 Matvei joined a physics circle at Kiev University; his practical-minded brother joined courses of stenography.

This is a good place to describe how Isidor shaped his life. He did a lot to make this book possible but he died in 1984 before it appeared.

While a student he suggested some improvements to the stenographic system he was studying and, jointly with some of his teachers, wrote a textbook. In 1926 he entered the Kiev Institute of National Economy with the purpose of formalizing economic science. In 1930, upon graduation, he was admitted to the Kiev State Architectural Workshops as an economist where he remained until his retirement (with the exception of four years during the War of 1941–1945). He took part in preparing three general plans of Kiev development and wrote on the demographic and economic substantiation of town planning.

He was always convinced that his brother surpassed him where the creative and intellectual potentials were concerned. Still, his own outstanding intellectual qualities were quite obvious: he could read ten languages and had a good command of Ukrainian (in 1941 he translated a book into this language). His knowledge of physics and the history of physics was amazing – a passion he developed after his brother's death.

He retained these enviable abilities and this intellectual potential until a very old age: at 60 he made independent and mathematically sound research in the field of complex analysis. He collected some twenty thousand volumes on history, philology, physics and mathematics in his small room in a flat that he shared with another family.

Isidor was destined to live through many personal tragedies: his brother perished when he was only 30; his parents died in Astrakhan after sent away from Kiev, when the fascists were approaching. He was a bad stutterer (at one time Matvei picked it up from him), and his condition worsened as blow after blow fell on him. It was much easier for him to write than to speak. He had no family of his own. Throughout his life he remained a kind and considerate person who sometimes was childishly naive. He was very tactful but never avoided of speaking his mind when necessary.

The circle that Matvei joined in 1925 was called "a physical section of the Kiev student society of nature investigators". It was set up by a young physicist Petr Tartakovsky1. It was a kind of a colloquium that prepared students for independent research and made it possible to select the most promising young people. At the time when active social processes were bringing all types of people together, Matvei Bronstein, who had no formal secondary education, was accepted into the circle intended mainly for the students of Kiev University.

It was equally important that Tartakovsky headed the circle. One of his former pupils wrote about him in 1940

> Tartakovsky was always very attentive and considerate. His enthusiasm was catching; he was always an attentive and benevolent observer of our progress, and he always met half-way the students wishing to probe deeper into physics. This was obvious in everything down to the minutest detail... I should say that not every scientist is able or eager to work with young people, supervise their progress in science and channel it. I find it hard to imagine him without his students. His life was dedicated to teaching [240].

Bronstein was one of those who flourished under Tartakovsky's attention. He was fully aware of this and gave his teacher his due ten years later. In 1935 he wrote in his CV: "I started my studies of theoretical physics back in Kiev (before I entered the University) under Petr Tartakovsky" [103]. Tartakovsky made it possible for him to work in the university library reserved for the teaching staff. Matvei had a desk for his personal use, which was a luxury. Another luxury at that time was efficient central heating. Bronstein used this privilege to the fullest extent.

They discussed problems related to classical physics, such as the dynamics of the boomerang, and the topical problems of contemporary physics. Very soon Tartakovsky became convinced that Matvei was superior to many of his other students. Several months after he joined the circle, in January 1925, Bronstein sent his first scientific article "On a Consequence of the Light Quanta Hypothesis" to the *Zhurnal Russkogo fiziko-himicheskogo obshchestva* [1]. He deduced a dependence between the boundary of continuous $X$-ray spectrum and the radiation angle from the hypothesis about the photon structure of $X$-ray radiation and the laws of conservation in the interaction of electrons and the atoms of the anticathode. A discovery of this effect would have added another argument in favor of the light quanta, "Otherwise, we shall somewhat clarify the question of the limits of the theory of the light quanta as applied to $X$-rays". At that time the photon theory was being actively rejected by Bohr himself (who changed his stand following the 1925 Bothe–Geiger experiments). With his article young Bronstein plunged into the troubled depths of physical discussions.

His words about the "limits of the theory" that appeared in his first article proved to be the key to his further scientific work and his world-view as a whole.

It is important to note that in his first article Bronstein paid particular attention to the experiment's practicability. He supplied a quantitative assessment of the expected effect for several working regimes of the $X$-ray tube and discussed the observation conditions.

Both the subject and the combination of a theoretician and an experimenter betray Tartakovsky's influence. He was an active supporter and an ardent propagandist of quantum ideas. Even his first publication [269] dealt with quantum physics. His books *The Light Quanta* (1928)[2] and *Experimental Foundations of the Wave*

*Theory of Matter* (1932) promoted quantum ideas in the Soviet Union. Both of them rested on a solid experimental foundation that left no doubts about physics being an experimental science. At the same time, he provided an idea of the radical transformation in the theory caused by quantum physics. Indeed, this was a unique combination for the period when theoretician and experimenter had already become two different trades. Bronstein was lucky to have been introduced into science by this man. After all, a theoretician needs an acute feeling of balance between his theoretical flying and the ground of an experiment.

Bronstein was working really hard: two of his articles on the quantum theory of the interaction of $X$-ray radiation and matter appeared in 1925 in the famous *Zeitschrift für Physik*; three more articles were published in the next year. All of them showed his solid mathematical basis.

Very soon he became known among the physicists and astronomers in Kiev; he was admitted into the section of scientists at the Kiev department of the Union of Workers of Public Education. He received favorable comments about his work from the Director of the Kiev Observatory and heads of physical seminars[3]. Their references helped Bronstein enter Leningrad University.

In 1927 his articles [2–4] were mentioned by A. Goldman in his "Physics in the Ukraine on the Tenth Anniversary of the Soviet Ukraine"[4]: "One feels inclined to draw attention to the works by M. Bronstein (a young Kiev physicist who later moved into RSFSR). He discussed the problems of the continuous $X$-ray spectrum, the quantum theory of the Laue effect and the motion of the electrons around the stationary center of the field" [166].

The young physicist did move to the RSFSR, or more exactly, to Leningrad in 1926, probably following Tartakovsky's advice. After all, Kiev was a provincial city and Bronstein had moved beyond its narrow limits.

# Chapter 2
# In Leningrad University (1926–1930)

## 2.1 Entering the University

Leningrad was the USSR's scientific capital, housing the Academy of Sciences and the main academic institutes until 1934. It was there that Bronstein became a physicist. Even though he was the author of several scientific papers, he still had to get a university diploma. One can imagine that. having been educated at home he was not over-enthusiastic about the prospect of studying according to official programs. No doubt he was aware of the gaps in his knowledge – after all, study is the major element in the theoretician's trade, yet at 19, Bronstein felt older than most of his fellow students. At the same time he could not but profit from being part of the physicists' community.

At that time, Leningrad boasted two higher educational establishments that offered a sound training in physics: the Polytechnical Institute with its Department of Physics and Mechanics and the university. The former, set up by Ioffe some seven year before, provided close ties between physics and technology. However, the training program included a wide range of engineering subjects and technological disciplines that a theoretician could have considered an unnecessary burden. Theoreticians were produced by the University despite it was a pretty old-fashioned. There were no famous scientists on its teaching staff (with the exception of D. Rozhdestvensky who was an experimenter). This was rooted in the past, when Petersburg University was inferior to Moscow University with its cluster of celebrities (A. Stoletov, A. Eikhenvald, P. Lebedev). Between 1907 and 1912 Paul Ehrenfest had managed to raise the level of Petersburg physics but in the twenties its teaching staff comprised mostly educators rather than researchers; Orest Khvolson (1852–1934) was one of them. He had written a definitive *Course of Physics* that was favorably accepted abroad (Einstein praised it and Fermi studied it). Though not young, he enthusiastically acclaimed the revolutionary theory of relativity and quantum physics.

Being a man of wide-flung interests, Bronstein was evidently attracted by the variety of subjects taught under one university roof: astronomy and philology, history and mathematics.

In 1926 Bronstein passed the entrance exams,[1] which probably presented no difficulties. Very soon he became a local celebrity with scientific papers in the best European physical journals to his name. His teachers were cautious not to become entangled in scientific discussions with him. He was known for his ability to pass any exam without much difficulty: early in November he went to Professor Khvolson to sit for an exam in general physics. Khvolson responded with: "You can't be serious, dear sir. The other day I read your article in *Zeitschrift für Physik*. It's not for you to sit for an exam in physics! Give me your record book". A week later he passed his exam in mathematics for the first year. His record book contained also signatures of V. Bursian, Y. Krutkov, P. Lukirsky, V. Fock, V. Frederiks [100].

Though he obviously could have finished his studies earlier he spent four years at the university. It seems that he considered the university a favorable milieu. Studies were by no means his main occupation during these years: it was at this time that he obtained some significant results in astrophysics that later (when academic degrees were reintroduced) earned him a degree of candidate of science without the usual procedure of defending a thesis.

It is very important for a young theoretician to have friends with whom he can discuss his ideas: young scientists tend to form rather stable groups of similarly minded people. In the spring of 1927 Bronstein was lucky enough to stumble across one such group, which he immediately joined. Significantly enough, it was poetry, rather than physics, that helped him.

## 2.2. The Jazz-Band

The group Bronstein joined was the famous Jazz-band, formed around George Gamow, Dmitry Ivanenko and Lev Landau[2], better known by their nicknames of Jonny, Dymus and Dau. They were called "the three musketeers". Some physicists and philosophers found it hard to follow the rapidly unfolding developments in physics and got polite indifference and littler respect from the "three musketeers". They reciprocated by calling them a "Jazz-gang".

Bronstein joined this musketeer group easily and naturally and faithfully served its royal majesty, physics.

Here is how Lady Peierls (1908–1986) described Bronstein's first encounter with the Jazz-band in a letter she sent to us on March 9, 1984. (She was Zhenya Kanegisser before when she married Rudolf Peierls in 1931, a German physicist whom she met at the Odessa physical congress. He was knighted for his scientific achievements, part of which could be attributed to his charming, optimistic and intelligent wife. After the spring 1926, when she joined the Jazz-band, she wrote much of its poetry.)

> I'll do my best to describe everything I remember of Matvei Bronstein. I first met him in early spring 1927. There were puddles everywhere, sparrows were chirping in a warm wind. On emerging from a laboratory on the Vasiliev Island I quoted a line from Gumilev to a young man who happened to pass by. He was not tall, wore large glasses and had a head of fine nicely cropped dark hair. Unexpectedly, he responded with a longer quotation from the same poem. I was

delighted; we walked side by side to the University quoting our favorite poems. To my amazement Matvei recited *The Blue Star* by Gumilev which I had no chance of reading.
At the University I rushed to Dymus and Jonny to tell to them about my new friend who knew all our favorite poems by heart and could recite *The Blue Star*.
That was how Matvei joined the Jazz-band. We were putting out the *Physikalische Dummheiten* and read it at university seminars. In general, we sharpened our sense of humor on our teachers and at their expense. I should say that by that time Joe, Dymus and Dau had left all others far behind where physics was concerned. They explained to us all the new and amazing advances in quantum mechanics. Being a capable mathematician the Abbot (Matvei Bronstein) was able to catch up with them quickly.
I can visualize Matvei with his specs slipping down his nose. He was an exceptionally "civilized" and considerate person (a rare quality in a still very young man); he not only read a lot but also had the habit of thinking a lot. He accepted no compromises when his friends "misbehaved".
I cannot say who gave him his nickname, which suited him perfectly: he was benign in his skepticism, appreciated humor and was endowed with universal "understanding". He was exceptionally gifted.

This is an ample illustration of the intensity with which the Jazz-band treated life. They had a seminar of their own at which they heatedly discussed events in physics and everything else under the sun: ballet and poetry, Freudism and the relations of the sexes were subjected to scathing and dissecting theoretical analysis. Here is an illustration of the tense atmosphere in theoretical physics put into verse by Zhenya Kanegisser who used Gumilev's *The Captains* as her model:

You all are the paladins of the Green Temple,
All leading your way through de Broglie's waves
The Earl Frederiks and Georgy de Gamow,
Who questioned the ether with nothing to save,

Landau, Ivanenko, two boisterous brothers,
Krutkov, the indifferent CTP's head,
And Frenkel, the general of Röntgen army
Who made the electron dance and spin,

The brilliant Fock, Bursian, Finkelstein,
And tiniest crowds of studying youths,
You started your voyage to follow Einstein
Who taught you to scorn the traditional rules.

Though Heisenberg's theories weren't a triumph,
And Born's hard-earned laurels seemed withered a bit,
Yet Pauli's principle, Bose's statistics
Have long won your hearts, and your minds, and your wit.

The Nature is still enigmatic and hidden,
You still do not know all the secrets of light,
The nuclear laws still remain undiscovered,
And you are now trying to conquer the blight.

When reading your cleverest papers in *Zeitschrift*,
With all our problems becoming more vague,
The only delight is the thought that Bothe
Will give you all guys a proper spank.

Here some explanations are needed – Y. Krutkov, who headed the Cabinet of Theoretical Physics, was not an overenthusiastic chief; Bothe's and Geiger's experiments deprived Bohr's non-conservation hypothesis in the Compton scattering of any grounds. This hymn was probably written in 1926. It appeared in the *Physikalische Dummheiten* (Physical Absurdities) that ridiculed tempestuous events around physics, sometimes with friendly irony and often with biting sarcasm. The editors themselves were not exempt.

The same could be said about the Jazz-band's relationships with their elders. The very name was a gauntlet thrown to public opinion: in the Soviet Union jazz was accepted in the late twenties. One of the scientific papers [156] was written jointly by Gamow, Ivanenko and Landau to honor one of the Jazz-band girls in the Astoria restaurant (where the university students had their meals for a token sum due to the efforts of the Commission for Improving the Life of Students).

Despite a somewhat frivolous reason, the article deserves a closer look which we offer below. No matter how talented the Jazz-band leaders could not create a paper from a restaurant atmosphere; obviously it was based on everyday discussions and ideas voiced by the entire Band. They offered no formalized substantiation and, therefore, they could not regard the results as belonging completely to physics. In all other respects it was a physical article that reflected the contemporary state of fundamental theoretical physics and even offered some glimpses of the future. In Chapter 5 we shall discuss it in greater detail and wonder why Bronstein was not among the authors.

## 2.3. The Abbot and His Astronomer Friends

No wonder E. Peierls did not know who had given Bronstein his nickname of the Abbot: it was given by another group he was close with during his student years, a group of astronomers.

At Leningrad University, astronomers belonged to the department of mechanics and mathematics, rather than physics. This was rooted in the past, when theoretical astronomy rested on a single arm of celestial mechanics.

Throughout history, the relationship between physics, astronomy and mathematics went through different stages. In antiquity, "physics" was the name of all sciences about nature; the obvious regularity of astronomical phenomena was taken as the model of all natural laws. Being mathematically precise, this regularity was an abyss between celestial physics, guiding the heavens, and earthly physics, struggling to put some sense and order into the chaos of the phenomena on Earth.

In the Newtonian age, new physics, new mathematics and new astronomy were born; Mitya Bronstein was first introduced to this remarkable time through Flammarion's *Astronomy for Everyone*. Newtonian physics declared that its laws were valid for the entire Universe, with celestial mechanics – the natural base for theoretical astronomy – being nothing more than a particular case of physics. Later, however, celestial mechanics became merely a part of mathematics. While astronomy rested on celestial mechanics, the students of astronomy were natural members of the department of mathematics and mechanics. True, they had to describe the movements of celestial bodies as motion of material points rather than physical bodies.

In the mid-nineteenth century spectral analysis as applied to the study of stars changed the situation. However, the union of physics and astronomy was finally sealed by quantum theory, which decoded the enigma of spectral lines. This opened new vistas for astronomy – astrophysics came of age in the second half of the twenties [234].

Any physicist who closely followed advances in natural science could never have missed astrophysics' heyday. The tree of astronomy was blossoming with the flowers of physics; astronomical numbers were turning into physical numbers.

Bronstein was such a physicist – while in Kiev he was fascinated by astronomy. No wonder then that at the university he attended lectures on astronomy. He quickly made friends among the students of astronomy: V. Ambartsumyan and N. Kozyrev; he was also friendly with I. Kibel, I.A., who studied hydromechanics. They divided their time between the university and the Pulkovo Astronomical Observatory.

What was more, Bronstein introduced his physicist friends to astronomy. When writing about Ambartsumyan and Kozyrev's early papers on stellar atmospheres, D. Martynov recalled that both had been members "of a talented group of students that was formed at Leningrad University in the twenties. It also comprised M. Bronstein, G. Gamow, L. Landau and D. Ivanenko – a veritable constellation of the future stars of the first magnitude! Bronstein and Ivanenko used to come frequently to Pulkovo to take part in free discussions of widely varied questions of theoretical physics and astrophysics that later gave birth to some significant papers. With his jet-black hair, the reserved, balanced, highly logical and convincing Bronstein was a decided contrast to Ivanenko, who was spontaneous and noisy and who spoke easily and fluently. He obviously knew what he was talking about; his mind was always brimming with barely formulated ideas. At that time, Bronstein had recently solved several important questions of the theory of radiation transfer in the atmospheres of the Sun and stars, while Ivanenko and Ambartsumyan had finished several papers on mathematical physics and the physics of the nucleus. It was during that period that Landau touched on some astrophysical problems – the result was his paper on the possibility of superdense stars that appeared in 1932 [234, p. 440].

In fact, the change of generations in astronomy was even more dramatic than in physics: the older generation was struggling under a double impact of physics and the new relativist and quantum ideas physicists themselves were just getting used to. Important observational facts were discovered. In particular, it was established that the spiral nebulae were other galaxies. This made the astronomical picture of the universe much wider than had been earlier believed. It was primarily young people who were introducing young physical ideas into astronomy; naturally enough, the older generation was distrustful of the attempts of their younger colleagues "to determine the number of atoms above each square inch of the Sun's surface" [234, p. 439]. They were shaken by the avalanche of new facts and ideas.

One day there appeared a notice at the department of astronomy informing those interested that M. Bronstein would read a survey of works by Bodichiraka Ramasatva, a prominent Indian physicist and astrophysicist, who, on visit to Leningrad, had kindly submitted his unpublished paper to them.

The lecture hall was filled to capacity – Bronstein had already earned the reputation of a brilliant lecturer. Having presented the basic assumptions, he formulated a problem for a planetary system's proper values. He chalked an imposing differential equation that contained Planck's constant, the velocity of light, the electron mass, the mass of the central luminary and a cluster of Latin and Greek letters. Bronstein discussed the wave function behavior and presented a range of proper values. Then he made certain transformations and inserted the mass of the Sun. At this point the audience recognized the famous Titius-Bode relation that determined the actual mean distance of a planet from the Sun. This was the main conclusion of the paper.

The audience was duly impressed. Speaking on behalf of it, Professor P. Gorshkov voiced his favorable opinion of "this extremely interesting paper" and offered his opinion on certain points.

The mysterious presentation was a complete success: it was revealed to the audience's laughter and to the great delight of the practical jokers.

The paper was read at the Astronomical Cabinet, the usual place for all lectures and discussions in astronomy. In 1927–1928 the students of astronomy put out a journal predictably called *Astrocabical Journal* that was very much like its cousin *Physikalische Dummheiten*. The titles reflected the state of affairs in contemporary physics dominated by the Germans and contemporary astronomy dominated by the British.

V. Ambartsumyan quoted a sonnet Bronstein had dedicated to the *Astrocabical Journal:*

I wish you be above critiques and praises,
I wish you be a beacon in the dark,
And like the sun you shine in all your phrases,
I wish you all the best, my newly born *Zhurnal*!

I know that circulation is but tiny;
Your fate was thought to be a lucky one
When maniacal Kostinsky3 was trying
To get and copy you for his scientific scum.

Your origin is rather enigmatic,
The efforts to reveal them are pathetic.

Your parents are unknown to everybody,
Your secret guards are mystery to all.
And once appointed they are always silent
Like priests of Dionysia of the old.

The young Leningrad astronomers were no less fond of nicknames than the physicists. More likely than not, these nicknames were derived from first names or family names: Ambarts for Ambartsumyan, Kib, Dau, Jonny, Dymus. Bronstein's nickname was of a different origin.

It was taken from a book by Anatole France *At the Sign of the Reine Redauque*, translated into Russian by I. Mandelstam, Zhenya and Nina Kanegisser's stepfather; their home was the favorite haunt of young physicists and astronomers.

The young astrophysicists read the book out aloud while commuting to Pulkovo; evidently they were greatly impressed by Abbot Jêrome Coiquard, doctor of theology and magister of sciences, and found many of his remarkable features in Bronstein: profound mind, wide knowledge, balanced skepticism, kindheartedness and tolerance. According to Ambartsumyan, Kozyrev was the first to apply this name to Bronstein, the scope of whose knowledge struck them most.

The abbot's devoted pupil was convinced that "no geometers and philosophers who, emulating M. des Cartes, were able to measure and weight the worlds could rival [his] teacher in talent and knowledge". It seems that Bronstein's friends found in him a true rival for the abbot, since in the twentieth century it was a privilege of physicists to measure and weigh the world.

Just like the fictional character Bronstein could not leave a book unread – he had read an awful lot of them. While it took Coiquard 51 years "to read all the Greek and Roman authors graced by time and human ignorance" it took Bronstein only 21 years to earn this honor.

As could be expected, a French theologian of the early Enlightenment and a Soviet physicist of the early socialist period were not identical: while the abbot was fond of his bottle, his food and other earthly delights, Bronstein was much more moderate – hence all contradictory explanations of how he got his nickname.

One should not imagine, however, that Matvei was engrossed in books and science to the exclusion of everything else. He and his friends were very much like other young people. Numerous photos of that time bear witness to this. In one of them the Abbot is holding a large cross. He is obviously converting Zhenya, who is kneeling in front of him; Ambartsumyan is nearby with a suitcase. It seems that he was posing as one of the traders who invariably followed missionaries to newly converted countries. In another photo there is a bespectacled young lady, modestly covered with a shawl, flanked by two young men. The "lady" is Matvei, and his two sweethearts are the Kanegisser sisters. Another photo was taken in Odessa during the 1930 Physical Congress. Young physicists in bathing trunks are holding laughing girls in swimsuits by their heels.

Late in the summer of 1929 Ambartsumyan, Bronstein, Kozyrev and Kibel travelled across Armenia to Ambartsumyan's home village. It took them slightly more than a week, during which they passed through a storm on Lake Sevan with waves of oceanic dimensions, rode in the mountains, spent a night in the open with horror stories told by turn and walked forty kilometers on foot. Being not very strong Bronstein had to mobilize all his inner resources, yet the main goal was attained: they distracted themselves completely from their intensive studies and scientific research.

## 2.4. The First Works in Astrophysics, Geophysics and Popular Science

Bronstein worked a lot in 1929 and achieved a lot: he wrote two papers on astrophysics, one on geophysics, his first popular science book and several articles. Indeed, one cannot expect more of a student!

His first astrophysical papers dealt with stellar atmosphere. Ambartsumyan and Kozyrev were working in the same field; this was the period when physics discovered a totally new object – the star as an integral physical system. To solve the main riddle – the star's internal structure and its energy source – one has to form an idea about its surface and the atmosphere that connects it with the outer world and the observer as a part. Without this, no in-depth studies were possible. On the other hand, while physics itself had advanced to the point where it could tackle the problems of the atmosphere, it had not advanced enough to look inside the stars.

The theory of stellar atmosphere had been developed enough to allow easy success for a chance intruder. It boasted of own masters such as K. Schwarzschild, J. Jeans, A. Eddington and E. Milne.

The problem of the radiation equilibrium of stellar atmosphere goes back to Schwarzschild. Astrophysics define stars (the Sun included) by the effective temperature $T_{eff}$, the temperature of the black body of the same dimensions and the full radiation equal to any given star. Its value is calculated by observations on Earth. Bronstein set himself the task of defining the dependence of the temperature of the stellar matter on the (optical) depth $\tau$ within the framework of the star's definite physical model. It had been established by that time that the dependence was

$$T_{\text{eff}}[3/4(\tau + q(\tau))]^{1/4}$$

where the value of $q(\tau)$ differed little and was established through the solution of a definite integral equation (Milne's equation). It was obvious that the numerical value $q(0)$ allowed one to determine the exact temperature of the solar surface $T_0$, through the value of $T_{\text{eff}}$ that could be measured on Earth. The best minds in astrophysics coped unsuccessfully to determine the exact value of $q(0)$. The result was several approximations – Jeans and Eddington produced two each and three belonged to Milne. It was Bronstein who in 1929 offered the exact value for $q(0) = 3^{-1/2}$ and, consequently, the exact correlation

$$T_0 = \left(\frac{\sqrt{3}}{4}\right)^{1/4} T_{\text{eff}}$$

(This result became known as the Hopf-Bronstein correlation [297, pp. 85, 96] although the order of the names should have been different since Hopf arrived at the same result later.[4])

The high level of Bronstein's first astrophysical papers is attested to by the fact that they appeared in major scientific journals. The third (and last) article on the stellar atmospheres was published in the *Monthly Notices* (Great Britain) on Milne's recommendation. It was an answer to a letter from Milne. It seems that he was greatly impressed by Bronstein's exact result and hastened to toss him another challenge – the boundary value of $q(\infty)$ (the infinite optical depth in the star's atmosphere corresponds to an insignificant actual geometrical depth). No exact result was obtained (and it has not been obtained so far), but Bronstein was able to produce certain approximations.[5]

These papers belonged to mathematical physics; he skillfully applied mathematics to resolve the already posed physical problems: there mathematics was not involved at the expense of physics. (Practically the same mathematical apparatus was invoked in the late thirties and forties to describe the transfer of neutrons in uranium).

We shall not discuss these works in detail here: every researcher is aware that time is harder on the creations of scientists than on artistic creations. This is especially true of theoretical physics. Even the most revolutionary ideas and works are preserved for posterity only as several lines and formulae in textbooks and definitve monographs. One or two sentences designed to educate a new generation or to express the author's emotional attitude to the results sum up a long and torturous path, painful efforts to overcome real and imagined obstacles, delusions and errors. Yet smallernothing remains of the uninterrupted flow of good or even excellent works. Only historian of science knows that they are needed to set up a favourable environment outside which no spectacular achievements are possible.

The high level of Bronstein's contribution to astrophysics was demonstrated by the 1934 granting of the newly introduced academic degree of Candidate of Sciences bypassing the usual procedure of defending the thesis. He also wrote articles on white dwarfs and on the influence of electron-positron pairs in the thermal equilibrium under high (stellar) energy densities (see Section 5.2).

In 1929 Bronstein turned to geophysics. Although the title of his large article [8] also contained the word "atmosphere", like his works in astrophysics this was a purely linguistic coincidence since the stellar and earth atmospheres are completely different spheres of research. The different physical conditions in them posed different physical problems: anybody probing the stellar atmosphere had to study the mean steady-state conditions that determined the star's temperature. The most important problems of the Earth's atmosphere are connected with its dynamics. Eddington was right when he said that a star was basically an simple research object that posed fewer riddles than man. One could say that the atmosphere of the Earth compares with man where its complexity is concerned: it is not for nothing that weather forecasts, supposedly based on atmospheric dynamics so far remain unreliable. It is as difficult to forecast weather as it is to guess how a man will behave under specific circumstances.

Bronstein introduced his article on the atmospheric dynamic with an epigraph from E. Kummer: "A certain degree of approximation can make a cobblestone an ellipsoid". This was a natural reaction of a physicist-theorist to theoretical geophysics. In general, theory can be applied only to comparatively simple models while the Earth, the main geophysical object, is far removed from the geophysical theoretical models – farther than is allowed in theoretical physics.

Bronstein was far from condescending to geophysics. In fact, his first popular science book, *Composition and Structure of the Earth*, is a good example of his profound knowledge of geochemistry, geophysics and seismology, unexpected in a specialist in theoretical physics, who one year later produced an article on quantization in the magnetic field and a detailed cosmological survey. In his book he presented vast observational material and discussed hypotheses that had nothing in common with theoretical physics, such as Wegener's hypothesis on continental drift.

One cannot but wonder how he was able to combine these far-flung fields – astrophysics, geophysics and fundamental physics. His varied scientific interests amazed his friends as well.

One reason for this can be found in his personal files, which show that in July 1929, while a student, he was working as a physicist in the Main Geophysical Observatory (MGO) in the department of theoretical meteorology under L. Keller

(1863–1939), one of the closest associates of A. Friedmann. He also studied the theory of atmospheric circulations. Bronstein's article [8] and some of his papers read at a seminar in MGO [209, p. 74] were related to this subject.

Who introduced him to geophysics? First, I. Kibel who worked in the same department in MGO and for several months had been doing his post-graduate work under Friedmann. He was studying the hydrodynamics of compressable liquids, or the dynamics of the atmosphere, these are closely related subjects. Their shared interests in the Earth's atmosphere were not disrupted by the fact that both tried to court the same girl (who rejected them both).

## 2.5. At the Shenroks on Vasiliev Island

There was one more man who could have introduced Bronstein to geophysics, and to fields far removed from theoretical physics. This was Alexander Shenrok who had come to the MGO (the Main Physical Observatory up until 1923) back in the last century. He was a pure meteorologist in the sense the word had in the nineteenth century, that is, he mostly observed weather changes. Throughout his student years Bronstein rented a room in Shenrok's flat on Vasiliev Island.

A German from Estonia, Shenrok studied in Germany (the scars on his face bore witness to his violent student days); after many years in St. Petersburg he became completely "Russified". In the post-revolutionary years he had to rent out part of his large apartment to alleviate the housing crisis. He was fond of university students, probably because he had no children of his own. Bronstein's friend, S. Reiser,[7] a philology student, also rented a room in the same flat.

They had first met in 1924 in the reference library of Kiev University; Reiser was allowed to use it having proved his worth at a seminar. He immediately noticed a dark-haired young man always immersed in books and journals teeming with formulae; from time to time he would write similar formulae on a sheet of paper. Very soon these two became friends: never in his life did Bronstein look down at the humanities. In this respect he differed greatly from Landau, who used the word "philology" to demonstrate his contempt for an inadequate physical paper and who believed that "philology was an occupation unworthy of a thinking man, something akin to collecting butterflies". In Leningrad, Reiser found himself in Eikhenbaum's circle; it was through him that Bronstein was exposed to the developments in literary studies that were blossoming at the time. He was always eager to plunge into the new books his friend brought back home. His photographic memory allowed him to memorize the contents and the layout.

In 1929, when M. Aronson and S. Reiser published their *Literary Salons and Groups* (edited by Eikhenbaum) Bronstein immediately greeted it with an ironic poem that ridiculed the then fashionable "montage method" that mostly relied on scissors and glue rather then on literary comments. Here Bronstein imitated Mayakovsky's peculiar style:

Before
  I thought
    that books are made this way –
They sit,
  and think,
    and wear off the trousers,
And many years elapse,
before this simpleton
Will taste the juicy fruit of his hard-working hours.
But seemingly
  Today
    This work becomes quite light:
It was some time ago
    That Reiser and Aronson
Have treaded path to praise
With scissors and some paste.

A great lover of poetry, Bronstein could recite many poems from memory, his favorite poets being Pushkin and Blok. Judging by his dedication pages, he knew a lot of German, English and French poetry. He was convinced that an ability to write verse was part of general culture. He himself wrote several poems that he never treated seriously. Reiser recalls, how in 1927, Bronstein showed him a small dark-green phial that supposedly contained cyanide he had procured from a chemist friend. When asked why he needed it, he smiled and answered with a poem with the Byronic title of "Euthanasia" of which Reiser remembered the following lines

Never, never I will be wounded
Never, never I will be in glee
In my waistcoat pocket always with me
Is the bottle that will make me free.
And the ghosts of the past dreadful years
. . . . . . . . . . . . . . . . . . . . . . .
In this world I am not a prisoner,
Nor a slave in the time-worn chains,
I am coated in iron warrior,
Smiling scornfully at my pains.
I am strong, I am calm, I am confident,
I am free to invite my death.
If the foe turns out more powerful,
By my choice I can draw my last breath.
I will not have to follow the conqueror
With a rope tied round my neck ...

In fact, Bronstein demonstrated his literary talent in popular science as well. We have already mentioned his first attempt at this, *Composition and Structure of the Earth*, which appeared in 1929. It followed the well-trodden path of an enlightened writer sharing his knowledge with a layman. The tone was neutral; the author's enthusiasm becoming obvious in the last paragraphs

> The lack of space does not permit me to discuss other thought-provoking ideas of the composition and structure of the Earth. I would like very much to have a closer look at living matter's role in the history of the Earth's crust, underestimated in the past. Today it is studied by geochemists such as academician V. Vernadsky, who writes about the biosphere, that is, about those of the Earth's mantles where life is going on.
>
> I shall not discuss this role here though it is very important for us if we want to know more about the celestial body we are destined to live and die on. So far mankind is chained to a small planet travelling in space around the extinguishing sun. People are urged to know more about the globe that serves them as home and, may be, eternal prison. But this may be incorrect: several centuries after Columbus set to the high seas to discover the fabulous riches of the New World, interplanetary rockets will probably carry new courageous conquerors into outer space. When the Sun exhausts its resources, mankind will probably unfold a banner of life on some other planet, under a brighter sun and bluer sky. This will add new meaning to geophysics and geochemistry studying the small planet – mankind's toehold for its plunge into infiniteness [55].

On April 4, 1929, he presented a copy to Reiser with a dedication that, in contrast to the high style of this passage, carried a good deal of self-irony.

His literary talent and profound knowledge enabled him to work quickly. He made a jocular dedication in his second popular-science booklet *The Structure of Atom* when he presented it to Reiser: "Labor productivity 24 pages a day. Royalties 301 r. 50 kop. To dear Monya in memory of the hard winter of 1929–1930". In fact, his intensive intellectual life was in contrast with relative poverty at home. He got no grants; the money he received from his parents was barely enough to cover his basic needs.

This was a period of heated debates on the theory of relativity and quantum mechanics. Matvei Bronstein learned from his own experience that real life and philosophy differed greatly where the predominance of the material over the spiritual was concerned. Like many other students he was obviously hungry, and the Shenroks used to invite him (like some of their other tenants) for dinner. Probably it was his host who introduced him to geophysics.

The winters presented the greatest problem of all – heating the rooms was the tenants' concern. It was extremely hard, if not impossible, to heat them. Before the revolution that was some twelve years away the flat belonged to L. Kasso, a notorious czarist minister of education. It was rather amusing to imagine him sitting on a small corner sofa planning police measures against students, the very sofa on which Monya Reiser slept. The students had to steal wooden scaffolding from a nearby construction site for fuel, yet they were not enough to heat the spacious rooms. Another remedy was to climb into bed under a heap of everything that would conserve heat and plunge into debates about everything under the sun.

Life in Leningrad was expensive: books and tickets for concerts and theatres without which life was impossible cost a fortune. Yet money was not the only thing that drove Bronstein to writing. He had an internal urge to explain difficult things and to lay bare the course of scientific thought. This was the time when popular science journal mushroomed throughout the country. Society had regarded science and technology as omnipotent. The cult of knowledge was prominent. It was not accidental that the journal entitled "Knowledge Is Power" appeared at that time at that time (in 1926). In 1929 Bronstein published a popular account of a paper by Einstein in which the great physicist made an attempt to combine gravitation and electromagnetism. He was an excellent guide in fundamental physics. His articles of 1929–1930 [54, 57–60] testify that he closely followed the developments in fundamental physics while being mostly engaged in astro- and geophysics. They also explain why Yakov Frenkel, head of the theoretical department of the Leningrad Physicotechnical Institute, annotated Bronstein's application for a job with the following words: "Bronstein is an exceptionally talented theoretician with wide-flung interests and profound knowledge. He shows much initiative and independence in everything he is doing. There is no doubt that he will be one of the best researchers in the department ." [284, p. 210]. Bronstein was 23.

# Chapter 3
# At the Leningrad Physicotechnical Institute

To better visualize the situation to which Matvei Bronstein was exposed upon his graduation, let's cast a glance at Leningrad theoretical physics and the Physicotechnical Institute.

## 3.1. Theoretical Physics in St. Petersburg and Petrograd

In the early twentieth century there was no theoretical physics in St. Petersburg to speak of – in general it was virtually non-existent as a separate field, in the way we are aware of it today. Even the giants of nineteenth-century physics, Maxwell and Boltzmann, never dedicated their time only to pure theory.

Paul Ehrenfest (1880–1933), a student of Boltzmann's, was one of the first theoreticians together with Planck, Einstein and Bohr. In Göttingen, where he moved upon graduation from Vienna University, he married a Russian girl, Tatiana Afanasyeva, a graduate of the natural science department of the Bestuzhev Courses. In 1907, they settled in St. Petersburg where Ehrenfest did a lot to promote theoretical physics in Russia [285].

He set up a Circle of New Physics that became an instant attraction for the students and lecturers of the university, the Polytechnic and the Electrical Physics Institute. It was the time of revolutionary changes in physics triggered by the quantum and relativistic ideas to which Ehrenfest actively contributed.

The circle allowed Petersburg physicists (A. Ioffe, Y. Krutkov, D. Rozhdestvensky and others) and mathematicians (S. Bernstein, Y. Tamarkin, A. Friedmann and others) to present their ideas and research results. Besides being a prominent physicist, Ehrenfest was a born teacher, able to assess any new theory critically: something that Einstein, Bohr and Pauli appreciated very much. The circle survived Ehrenfest's departure for Leiden in 1912, where he succeeded Lorentz.

By the mid-1910s the Polytechnic had emerged as an experimental center of new physics: A. Ioffe, who had studied under Röntgen and was a member of Ehrenfest's circle, launched his own experimental studies in the laboratory headed by Prof. V. Skobeltsyn. He also lectured at the University and proved to be a magnet of attraction for numerous students and post-graduates: Y. Dorfman, P. Kapitza, P. Lukirsky, N. Semenov, Y. Frenkel and others. The Seminar of New Physics at the Polytechnic preserved Ehrenfest's traditions: surveys of the latest achievements in physics and presentations of new ideas and results. Frenkel was the only theoretician among them; at the 1916–1917 seminar he presented his papers on classical electrodynamics and the electron theory.

The World War and the economic and cultural isolation of Russia that followed the 1917 revolution were not conducive to the advancement of science. Economic disruption made experimental research practically impossible; no scientific journals arrived from abroad; there were little if any contacts between scientists from different cities; there was no question of international contacts. Yet, a radical turn in Russian physics took place during these grim years.

An institute of physics and biophysics was set up in Moscow in 1918 by academician P. Lazarev, a pupil of P. Lebedev. In Petrograd, Ioffe, Nemenov and Rozhdestvensky set up new institutes. The Soviet government sponsored the State Institute of X-Rays and Radiology and the State Institute of Optics (GOI); very soon the former split into the Institute of X-Ray (and medical biology), the Institute of Radium (headed by V. Vernadsky) and the Physicotechnical Institute (FTI) under Ioffe.

The FTI department of theoretical physics was very small: it consisted of Bursian, Frenkel and students of the newly organized Department of Physics and Mechanics.

Krutkov was the GOI leading theoretician. The institute maintained close ties with the university in the form of joint seminars of the Institute of Physics at the university and the GOI. Young V. Fock and Professor V. Frederiks became involved in the latter half of the twenties. Theoreticians were at a premium and worked in both institutes at one and the same time.

## 3.2. The Physicotechnical Institute and Its Seminars

Bronstein came to the FTI in May 1930. After ten years of rapid growth, the administration, supported by the government [265], decided to spare its leading scientists to set up new physical institutes. Back in 1927 the Institute of Heat broke away from the FTI. P. Tartakovsky arrived in Tomsk (Siberia) in 1928 to work in the Siberian branch. In 1929 a Ukrainian Physicotechnical Institute was opened in Kharkov. It was staffed with leading research associates from the Leningrad mother branch, headed by I. Obreimov, Ioffe's deputy. This was a self-imposed brain-drain to solve a task of national importance – to set up scientific centers throughout the country. Ioffe believed that efficient research and management were impossible at a large institute. In 1931 it was decided, therefore, to transform two large sectors headed by N. Semenov and A. Chernyshov, into the Institute of Chemical Physics and the Institute of Electrical Physics.

Early in 1930 the FTI consisted of several sectors, each sector divided into groups, while the groups in their turn were divided into teams. Here is a list of groups, their heads and the number of associates in Ioffe's sector of physics and mechanics:

Group 1. The problems of energy (A. Ioffe, 28 associates): teams of thermal electrical phenomena; photoelectrical phenomena; geliotechnology; the sources and receptors of short waves.

Group 2. Physics of crystals (I. Kurchatov, 17 associates): teams of ferroelectrics, liquid crystals, the physics of ice and crystal formation.

Group 3. The physics of metals (Y. Dorfman, 43 associates): teams of phase transformations, plastic deformations, the properties of metals under dynamic loads, the role of free electrons, the magnetic properties of surface layers.

Group 4. Biophysics (G. Frank, 4 associates).

Group 5. The conditions of emission of X-rays and electrons (P. Lukirsky, 10 associates): teams of the mechanism of X-rays and their effect on atoms and electrons, the nature of electron emission.

Group 6. Theoretical physics (Y. Frenkel. It consisted of (quoted in full from a document from the Institute's archives): "Team 1 – theoretical physics, head V. Bursian, senior engineer L. Landau, engineers V. Fock, V. Kravtsov, A. Samoilovich, B. Davydov, A. Timoreva, scientific associate G. Mandel, senior engineer M. Bronstein [his name was added in pencil]. Team 2 – mathematical physics, head M. Machinsky, engineer A. Markov, engineer P. Artemov".

There was also a group of methodology headed by L. Rubanovsky, with 4 associates.

The staff comprised 220 associates, 80 of whom were engaged on technical jobs.

In 1930 the Theoretical Department employed 13 people. There was another theoretical department at the Institute of Chemical Physics (not yet detached from the mother institute) headed also by Frenkel. He had L. Gurevich, S. Izmailov, M. Elyashevich and O. Todes in his department; Y. Zeldovich started along his road of physics there as well. G. Grinberg was working at the Electrical Physical Institute, while Yu. Krutkov and G. Gamow were engaged at the university and the Institute of Optics.

These were the theoreticians who worked in Leningrad. Moscow could hardly boast more theoreticians. In fact, early in the thirties theoretical physicists were as unique as astronauts today.

At all times, theoreticians needed close cooperation, the seminars being the most efficient form of it. The FTI ran seminars according to its problem range. There were two permanent seminars (general and theoretical) and monthly meetings of the Learned Council that discussed scientific as well as administrative problems. Five years after he came to the institute (on May 16, 1936), Bronstein was appointed a council member.

There were also group councils and thematic seminars, some of them being merely *ad hoc* structures. Others, such as the nuclear seminar and the seminars on solids, electrical phenomena, liquid crystals and mechanical properties, met regularly and were functioning for a long time.

They discussed recent developments in their respective fields as summarized by their members, provided consultations for experimenters, supplied calculations and dealt with special theoretical problems. In 1936, Bronstein was transferred from the semiconductor laboratory to the nuclear seminar headed by I. Kurchatov and A. Alikhanov. Seminars figured prominently in the list of duties of the institute theoreticians; according to the order issued in 1936, they had to "(1) attend the general seminar; (2) attend the theoretical seminar; (3) submit their personal timetable of theoretical consultations for the laboratories they were attached to" [104].

The order itself was prompted by the fact that the theoreticians engrossed in theoretical studies tended to neglect the laboratories. Some hotheads even suggested the theoretical department be disbanded and its associates affixed to the laboratories. The administration decided against such a radical move.

In the 1931 research program, Bronstein was listed as studying anomalous phenomena in dielectrics together with Kurchatov. At the same time he was responsible for the studies of the theory of radiation balance in stars and nebulae.

Late in 1935 the Learned Council discussed the report on the institute's work designed for the famous 1936 March session of the Academy of Sciences. Frenkel and Bronstein had to prepare the section dealing with theoretical physics. They pointed to the special role the Institute's conferences played in the development of physics, discussed the philosophical implications of physics and looked into the situation of book publishing and scientific periodicals. By that time Bronstein had earned himself the reputation of a solid researcher who was invariably invited to speak at the defense of theses, supervise admission of post-graduate students; his opinion about the Institute's performance was always sought afterward.

The seminar, headed by Frenkel, had launched its weekly sessions in the mid-twenties. It was mainly attended by research associates from the FTI and students of the Department of Physics and Mechanics. Physicists from other institutes and the university were also attracted. At first there were ten people who came regularly: ten years later there were twenty. Formally a Leningrad seminar, it accepted guests from other cities – Tamm used to come down from Moscow, Landau from Kharkov, other people from Kiev, Sverdlovsk and Odessa. It also proved to be a point of attraction for many prominent physicists from other countries: Bohr, Born, Dirac, Langevin, F. London, Mott, Peierls, Pauli and others.

Its national and international success can be explained by a friendly and businesslike atmosphere – criticism was constructive and questions to the point.

Though no mean mathematician himself Frenkel spread formulae economically, skipping the intermediate stages. He cut quickly to impressive results through extremely simple constructs: it was his conviction that this road was open to anybody diligent enough to look for it. Here are his comments on a thesis: "Calculations are extremely complex. One cannot but marvel at the author's patience as he was cutting a road through a forest of equations .... For my part I would have been discouraged and looked for a simpler way: it takes much more than diligence to prop science further – sometimes laziness can be just as useful. The author should have looked for a shorter way to the same results – what he needed was an asymptotic result that could be easy to trace back" [284, p. 436].

It seems that the young theoreticians (Bronstein included) did not share this conviction (sometimes "mathematics may outwit man"); at the same time it did them a lot of good to watch this purely physical approach.

The seminars embraced the whole of physics: semiconductors, solids and liquid bodies, phase transitions, nuclear physics, magnetism. Throughout their history the participants heard hundreds of papers and tens of speakers. It seems that Bronstein left an especially deep impression on all of them. He submitted an interesting paper on Pauli's contribution on magnetism to the 1930 Solvay Congress. At another sitting he contributed a series of summaries on Wilson's works on semiconductors and a report on his own results in the same field. In the mid-thirties, when the theoreticians of the FTI turned to nuclear physics, Bronstein was attached to the department of nuclear physics. He spoke at Kurchatov's seminar and supplied his survey on the experimental results obtained by Fermi's group on slow neutron nuclear reactions to the theoretical seminar.

Every Friday the general institute seminar met under Ioffe's supervision. It left an indelible imprint in the minds of many of its participants. One of them, V. Berestetsky, has even supplied a "science fiction essay" [134] in which Ioffe, Frenkel, Landau and Bronstein can be easily recognized.

Let us now go back to May 1, 1930, the day Bronstein joined the Leningrad FTI and look at his main scientific achievements. Frenkel's high assessment of Bronstein quoted in Chapter 2 was based on his knowledge of this young man's potential. May 1930 is the date when their joint work was finished [9].

## 3.3 "Quantizing Free Electrons in a Magnetic Field"

By that time non-relativistic quantum mechanics had already been formulated. The current task was to apply its principles to specific problems. Dirac's equation was the main achievement of relativistic quantum theory. Though it described the electron's magnetic moment (a remarkable achievement), other conclusions seemed to be paradoxical. One of them was made by Rabi in 1928 [256]. Based on the formal solution of Dirac's equation, he concluded that the energy of free electrons in a uniform magnetic field should be quantized; he did not extend his effort to an analysis of experimental consequences. Mathematically immaculate, today his work calls for no further verification. Back in 1928 the response was different. Thirty years later Landau wrote: "I can vividly recollect how, in 1930–31, all physicists, Dirac included, agreed that no matter how elegant, his theory was incorrect since it allowed for particles that could not exist. From the viewpoint of experiment its results were absurd" [217]. (The existence of positrons was experimentally proven in 1932.)

An article by Frenkel and Bronstein stated in its very first paragraph: "It is important to demonstrate that quantization inevitably appears in any form of quantum theory – be it Bohr's 'semi-quantum' mechanics or the wave mechanics of Schrödinger and Dirac – to become convinced that the discrete energy levels of free electrons in a magnetic field is not one of the Dirac equation's paradoxes. It corresponds to a real physical phenomena that has not been experimentally proven".

By quantizing an electron's rotation in a magnetic field with the help of Bohr's postulate $mvr = n\hbar$, and then with the help of a "more correct" condition $[mv - (e/c)A]r = n\hbar$ ($A$ is a vector potential of magnetic field $H$) they got equispaced energy levels $W = n\hbar\omega_L$ and $W = 2n\hbar\omega_L$ ($\omega_L = eH/2mc$ being Larmor frequency).

The solution of the corresponding problem of quantum mechanics confirmed and specified the second formula for the spectrum $W = (2n + 1)\hbar\omega_L$, with the energy of the principal state being $W_0 = \hbar\omega_L$. By demonstrating that a wave of only one length $\lambda = \pi c/\omega_L = 10^4(\Gamma c/H)$ was possible in transitions the authors established the selection rules. This was the first reference to the resonance nature of interaction between quantum electrons and radiation. Today this process is prominent in the magnetic optics of solids (in the early fifties, experiments were staged to observe the cyclotron resonance in metals and semiconductors).

In 1930 this radiation could not be experimentally observed therefore the authors pointed to another phenomenon, "the tendency of free electrons to transfer spontaneously into the principal state with the minimal rotational energy $W_0$." They called it paradoxical and missed an opportunity to forecast a new phenomenon – electron diamagnetism in metals – that could be deduced from the spectrum they found.

Landau made this forecast in the paper [213] that appeared simultaneously with the Frenkel–Bronstein article. In the spring of 1930, while in Britain on a Rockefeller grant, Landau was stimulated by his discussions with Kapitza on the anomalous properties of the electric conductivity of bismuth in strong magnetic fields (the Kapitza effect). The result was Landau's theory of diamagnetism. In the first part he coped with the same task and, naturally, obtained the same energy spectrum.

It is interesting to note that four months before Rabi published his article, the same journal published a paper by V. Fock on quantizing the harmonic oscillator in a magnetic field. An oscillator with the $\omega = 0$ frequency was regarded as a free electron. The formulae for quantizing its energy in a magnetic field within the $\omega \to 0$

limit could be deduced from Fock's formulae for the oscillator. But this fact was passed unnoticed [253].

## 3.4 "A New Crisis in the Theory of Quanta"

The All-Union Physical Congress was Bronstein's first large forum. It was held in Odessa between August 19 and 24, 1930, by the Russian Physical Association. It attracted more than 800 people who read 200 papers on the entire range of physics. Sommerfeld, Pauli, Simon and Peierls were among its foreign participants.

It was an important event for Odessa, a port in Southern Ukraine. The city fathers lent their own assembly hall and the best hotels; the opening ceremony was broadcasted; the service bureau offered tickets to theatres, cinemas and sightseeing tours; the delegates were supplied with free travel cards. True, all of them preferred to spend their free time at the fabulous beaches.[1]

One can surmise that as the author of a paper on the quantization of free electrons in a magnetic field Bronstein would have been interested in the panel on the electron theory of metals. This panel heard the papers by A. Sommerfeld (on the magnetic field's influence on electric conductivity, I. Tamm and S. Shubin (on the selective photoeffect) and L. Shubnikov.

There is evidence, however, that Bronstein was mostly interested in "A New Crisis in the Theory of Quanta". That was the title of the article he wrote in August–September, 1930. It appeared in the *Nauchnoe slovo* journal [64].[2]

He outlined the evolution of physics and pointed to the fact that physics that went beyond the macro-world "could not be satisfied with the primitive experience of a savage or the narrow experience of a nursery; in short, macroscopic experience". He was quite eloquent when describing a "new" crisis in quantum physics when a relativistic quantum theory became an urgent necessity. (The "old" crisis was resolved in the mid-twenties when Bohr's theory was replaced with quantum mechanics.) Bronstein mentioned several of the manifestations of the new crisis: the $\pm$-difficulty of Dirac's equation (negative energy states), the infinite nature of electron energy, the mystery of the nucleus (the "electrons inside the nucleus" in the first place).

All these problems related to distances of order $\sim 10^{-13}$ cm, that is, the size of nuclei. Besides, there were considerations that a matter of this dimension could not in principle be exactly measured. It was quite reasonable to suggest, therefore, that to overcome these difficulties quantum mechanics should be transformed in such a way as to allow adequate reflection of the principal impossibility of measuring the lengths of "an inner electronic order". One can see that the author thought the situation over and was aware that his considerations were of a preliminary nature. Still, he wrote with evident sympathy that "there was a number of physicists –

Heisenberg from Leipzig, Ivanenko from Kharkov and Ambartsumyan from Pulkovo – who independently arrived at the idea of 'quantization of space', that is, a theory that would comprise a unit of the 'smallest possible length' such as 'an atom length' (lengths smaller than that would have been deprived of all meaning".

Clearly impressed by the discussions at the conference on quantum mechanics (June–July 1930, Kharkov), Bronstein discussed this idea at length in his article. Ambartsumyan and Ivanenko also sent their contribution from Kharkov [94]; it was dated July 21. They suggested replacing the normal continuous Euclidean space with a discrete entity of points that would form a cubic lattice that would look like an infinite crystal. The differential field equations were substituted for difference equations ($df/dx \to \Delta f/\Delta x$), the solutions of which included the lattice constant, allowing one to get rid of the infinite proper energy.

This gave rise to a fundamental difficulty – how to combine this lattice and obviously non-isotropic space with relativity theory.[3]

Ambartsumyan and H.D. Ursell, a young British mathematician, tried to remove the difficulty by finding probababilistic ties between observations in different reference systems, that is, to generalize Lorentz's transformations statistically. Bronstein discussed these efforts at the Odessa Congress.

He was not an impassioned eye-witness of the stormy events in the theory of lattice geometry. In a postcard he sent to Frenkel on August 9 from the Crimea, that is, after the Kharkov conference and before the Odessa Congress [284, p. 212], he wrote

> Dear Yakov Ilyich,
> Here is the house in which I am living. I cannot point out my window because it looks out on the other side.
> I am following your instructions and behaving myself: I go seabathing every day in a futile attempt to learn to swim, read Born-Jordan, Wintner (*Unendlichen Matrixen*) and detective stories borrowed from a Miskhor library and in general enjoy a peaceful life. I stopped translating Dirac since Dymus never sent Chapter 1 no matter how much I begged him. I checked Ambartsumyan's formulae on lattice theory and found them incorrect, and so on.
> I'm touched by your warm attention to my trousers though I should mention that in this warm climate people mostly wear shorts.
> Ioffe came here as well to "warm his chilling blood in the sun" (Pushkin).
> Best regards to Sarra Isaakovna,
> Yours, M. Bronstein.[4]

The problem of quantization of space was heatedly discussed on board a ship that took the congress participants across the Black Sea to Batumi. Some of them, like Pauli, had no hopes for quantum geometry; Bronstein quoted him as saying: "Those who are making holes in continuous space should mind where they step". This was what Pauli wrote in an editorial designed for the first issue of "Am Morgen nach der Schlacht", a newspaper Bronstein produced on August 26, 1930. It carried reports of the heated debates that lasted well into the night. (Here he called on his experience of putting out *Physikalische Dummheiten* and *Astrocabical Journals*.)

In the heat of a battle somebody quoted

*Die Esel fassen kaum es*
*Die Quantelung des Raumes.*[5]

In his conclusion of the "new crisis" in quantum theory Bronstein pointed to a general trend in scientific advances toward eliminating some ideas inherited from classical physics: "The real world might prove to be at variance with our ideas no matter how indispensable we think them to be." He then quoted Heisenberg as saying that quantum electrodynamics sinned when it applied to the microworld Maxwell equations and the concept of the field based on the classical ideas of motion belonging to the macroworld. This statement, formalized by Landau and Peierls [221], was impressive enough till 1933, when Bohr and Rosenfeld neutralized it by their analysis (for more details see Chapter 5).

Heisenberg's article of August 1930 bears traces of his attempt to develop discrete geometry. He wrote about minimal length and difference equations and supplied, at the same time, a very simple objection to the new discrete approach. He insisted that in the domain of relativism, where the rest-mass of electrons and protons may be ignored when compared to the particles' energies, quantum-relativistic theory should be based only on the fundamental constants $c$ and $h$ that could not be used to form a dimension length that could claim to be minimal. Bohr and Rosenfeld repeated this consideration in their 1933 work. Their careful analysis of the measurement procedure allowed in quantum electrodynamics reinstituted the "field in a point" concept over which Landau and Peierls had cast doubt.

It is known today that the gravitational constant $G$, the third universal constant, should be taken into account along with $c$ and $\hbar$. Back in the thirties everybody believed that gravity was walled off from quantum physics by considerations of the practical and quantitative nature. At least, this was what Heisenberg and Bohr were preaching; not all their colleagues shared their delusions: some of them, like Einstein, went to another extreme of overestimating the $c$ and $G$ constants. As always, the truth was somewhere in the middle. Bronstein seemed to be closer to it in 1935 when he (the first among his colleagues) involved all three universal constants into a physical analysis. He was able to predict an inevitable and radical reconstructing of the physical picture of the world in the *$cG\hbar$-domain*. We shall discuss this in detail in Chapter 5.

In 1931 he finished his article with the words: "Most of the theoreticians seem to be at a loss when confronted with still unresolved and seemingly unsolvable problems. This is one of the typical traits of the current crisis."

Indeed, this feeling was deep enough to drive some physicists to the idea (formulated by Bohr) that the future reconstructing would sacrifice even the law of the conservation of energy (for more details see Chapter 4). They still remained under the spell of the preceding thirty revolutionary years.

Pauli was critical of both discrete geometry and Bohr's hypothesis but held a similar view of the "new crisis". In 1933, when the fundamental difficulty of Dirac's equation triumphed with a forecast of anti-particles and Pauli's idea of neutrino prevailed over Bohr's awfully bold hypothesis, Pauli himself was still insisting that a genuinely quantum-relativistic theory would "lead to considerable changes in not only the field concept but also the space-time concept in the areas of dimensions $\hbar/mc$ and $\hbar/mc^2$" [249, p. 190]. This was the way the entire generation of physicists destined to live through the "new crisis" was thinking.

The hypothesis of minimal length, an offspring of the "new crisis" was an attempt to generalize geometry in a quantum-relativistic way. These attempts had a history of their own and started with the unified field theory of the twenties [127].[6] Here is what Bronstein had to say in connection with the next unified theory draft: "A geometry of space and time that would embrace the laws of gravitation and the electromagnetic field together with the quantum laws is the boldest challenge physics ever faced" [54]. Hence a burst of enthusiasm over space quantization.

He was of the opinion that "even if the program of discrete geometry fails it will leave its traces in physics" [64]. Indeed, the idea of quantum geometry, or, to put it more cautiously, the idea of the fundamental length (that limited classical Euclidean geometry's range), has not vanished [200]. At different times different people pinned their hopes on it: in the sixties Igor Tamm was among its enthusiastic supporters. There was a whole crop of quantum geometries – non-commuting coordinates, finite geometries, curved momentum space, etc. All of them associated fundamental length with $10^{-13}$, the same unit that had been suggested in 1930. When tested for smaller distances quantum electrodynamics proved to cede its ground to Euclidean geometry for at least several orders.

For a long time attempts to generalize the space-time description were dominated by elementary particle physics in the sense of the word that did not take gravitation into account. It is believed that all this belongs in the past, yet, the idea that a generalized space-time description is inevitable is gaining supporters – there were few of them before the seventies. This description is seen as part of a unified theory including gravitation and quantum cosmology. This generalization is connected with the so-called Planck length $l_{Pl} = (\hbar G/c^3)^{1/2} \approx 10^{-33}$ cm. It was Bronstein who was first to discover the grounds for this forecast back in 1935 (more details in Chapter 5).

Bronstein opened his article on "the new crisis in the theory of quanta" with the story of a meeting of prominent experts in quantum physics who, being overwhelmed with quantum theory's problems, solemnly vowed to abandon it. The list was punctuated by Pauli blowing a horn. This happened in the spring of 1930 at Bohr's. On the whole, Bronstein's article is full of happy anticipation of future scientific advances rather than sour pessimism.

This optimism notwithstanding, the editors of *Nauchnoe slovo* refused any hints at a crisis in Soviet physics. This is why they forwarded the article with the words:

> while discussing one of the most acute crises of bourgeois theoretical thought in natural science – the crisis of theoretical physics – the author failed to refer to the crisis of the bourgeois idealist-Machist world-view. He ignores a possible exit from the theoretical dead-end through dialectical materialism.

The editors tried to justify themselves: "The author provides a vivid picture of contemporary quantum physics that even a layman can enjoy." To be safe rather than sorry, they supplied the article with an end sentence: "It is impossible to overcome the crisis with bourgeois science's internal means." As A. Anselm recalled, an enraged Bronstein had been planning a retribution little suited for a pure theoretician.

All this gives an idea of the social and scientific atmosphere of the twenties and thirties.

## 3.5 Science and Society

In the first decades of Soviet power, social supervisors were closely watching how natural science progressed. There were weighty reasons for this interest: the natural sciences that supplied new ideas to technology were regarded as powerful forces that could transform the productive basis and, according to Marxism, society as a whole. Besides, the revolution in natural science associated with relativism and quantum physics was seen as a process that paralleled the social revolution.

Relativity theory was accepted with special warmth – Einstein became a household name. Indeed, anybody who had to answer the questions of "when" and "where" in the course of every day could not but be impressed with relativity effects of space and time. People were also excited by the findings of the British astronomers who had confirmed the German physicist's theory – this seemed to be a miracle in a world recently torn apart by a war and nationalism. Nobody, not ethnographers nor poets nor theologians could resist the temptation to discuss the theory of relativity as they understood it; in the twenties popular of it accounts appeared by the dozens [130].

Naturally enough, relativity theory attracted many ardent opponents as well. According to Planck: "new scientific ideas triumph not because their opponents become converted but because they gradually die out" [254, p. 656]. Anybody born into the world after the theory of relativity has established itself will find it hard to believe that to a physicist of the prerelativist age its acceptance did not come easy. Henri Poincare himself who had done a lot to make relativity theory possible, sometimes voiced anti-relativist ideas.

Arkadi Timiryazev (1880–1955), professor of physics at Moscow University, headed the ranks of anti-relativists in Russia. Academician V. Mitkevich (1872–1951), an expert in electrical engineering, as well as quite a few physicists, philosophers and journalists sided with him. They armed themselves with all the physical and quasi-physical arguments they could master; to strengthen their position they stooped to non-physical or even anti-physical arguments, borrowing them from philosophy and politics.

In the late twenties the philosophical discussions around General Relativity had reached their peak.

The ranks of academics swelled with young men from the social strata previously deprived of the privilege of higher education; this sapped the older generation's authority and pushed Russian science onto the world level.

The new men were keenly aware of the boiling ideological atmosphere of the time and the social importance of science; this feeling has abated since then.

The quoted foreword to Bronstein's survey [64] was nothing exceptional – the same issue carried a wordy declaration signed by the young outstanding mathematicians L. Lyusternik, L. Shnirelman, A. Gelfond and L. Pontryagin that called for a radical reconstruction of the Moscow Mathematical Society. They wrote, in part, that "it should bring mathematicians and proletarians closer together to fight in serried ranks for the Marxist revolutionary approach to mathematics and to liberate Soviet science from bourgeois ideological captivity."

In 1935, *Izvestia* published an article by Lev Landau under the telling title "Bourgeoisie and Physics Today." He hotly accused Western physicists of succumbing to bourgeois influences; Eddington and Jeans were branded as "mediocrities whose scientific contribution does not merit attention", Bohr himself was not spared in the heat of the battle [215].

There was more than just the energy of the social transformations triggered by the revolution behind this. The development of physics – the theory of relativity and quantum physics – the radical reconstruction of the foundation and renovation of the entire building required more attention spent on the methodology and philosophy of physics. This was the hallmark of the twenties and thirties that could be detected in the *Uspekhi fizicheskikh nauk* editorials, in the fact that the FTI deemed it necessary to set up a methodological seminar and that at the Odessa Congress B. Hessen spoke at length about the methodology of quantum physics and the relationship between physics and philosophy.

At that time any physicist engaged in theoretical studies of quantum-relativistic physics was forced to ponder methodology. It was not in Bronstein's nature to avoid methodological challenges of the time. All his works, even those written at the very beginning of his career, offer profound observations and succinct definitions that greatly contributed to a better understanding of the methodological lessons physics taught its priests [178]. True to his nature, he did not avoid problems of methodology in his article on the "new crisis" [64].

Those who wrote the foreword were acting under the conviction that planned economy was the answer to all economic riddles. It was the second year of the severest recession in the West and the second year of the first five-year plan period in the Soviet Union. Everybody was under its spell – it was believed that planned development would accelerate scientific advances as well. Ioffe spoke on this subject at the Odessa Congress; in March 1931 the first All-Union Conference on Planning Scientific Research followed, in December 1932 came the second conference. One can easily imagine that anybody at home with the prevailing ideology of the time and but knowing little about fundamental science, its past and present developments, would be tempted to draw a parallel between the economic crisis in the West and "the new crisis in the theory of quanta". Obviously this line of reasoning would prompt a conclusion that Soviet science, armed with the revolutionary ideology and organized in accordance with the principles of planning, had nothing to do with the crisis and was able easily to overcome it. This was what inspired the foreword.

Bronstein was fully aware of the dialectics of scientific progress and saw advantages of its planned development. On the other hand, he would never have ignored the fact that unexpected discoveries (such as those made by Becquerel and Röntgen) and novel theoretical ideas (such as Planck's quantum or Einstein's geometrization of gravitation) would narrow down the possibilities of planning in science. Traces of this can be seen in a caricature showing Bronstein in gypsy woman attire pronouncing his verdict: "To plan is to forecast".

No matter how skeptical he was about telling the fortunes of fundamental physics, he was convinced that its further advance would rest on quantum mechanics and the theory of relativity.

## 3.6. Quantum Mechanics in the Early Thirties

Two reviews that appeared side by side in *Uspekhi fizicheskikh nauk* in 1931 provide insight into Bronstein's ideas about quantum mechanics. He offered his opinion on Dirac's *Principles of Quantum Mechanics* and Weyl's *The Theory of Groups and Quantum Mechanics*. Here is what he wrote:

> Several years of rapid development of quantum mechanics have come to an end – today it is rather well-ordered. The ideas that seemed abstract and extravagant are now as familiar as the back of

> one's hand. The limits to its applicability and the fundamental difficulties that prevent it from expanding beyond these limits are becoming increasingly clear – in short, we are facing a crisis. No wonder, therefore, that the theoreticians deem it necessary to look back at the road traversed, to sum up the accumulated experience and to analyze the basic principles. They need this to charter the way forward.

This was how he opened his review of Dirac's book, "the best exposition of quantum mechanics to have appeared so far." The book was far from ideal, however: Bronstein believed that Dirac did not pay sufficient attention to the uncertainty principle and underestimated the radical nature of the changes that would come out of the relativistic quantum theory.

He spoke about the shortcomings because they were "less obvious than its undisputable merits, the chief of them being the simplicity of presentation." He wrote that the chapters on the theory's specific applications could be used as a textbook, while Dirac had brilliantly managed to disprove "a myth about contemporary physics being a jungle of mathematical formulae: throughout the whole book the reader will never stub his toes against sham academism and silly pedantry."

This was an opinion worth listening to: Bronstein was not a superfluous reader. In 1932 he translated the book into Russian and supplied it with comments.

Weyl had less luck: while paying tribute to his knowledge the young reviewer continued

> The presentation that is as elegant as everything that Weyl writes is marred by pedantry. This is not a physical book but a mathematical composition about physics. One should not open it to acquaint oneself with quantum mechanics; it is aesthetic to the point where a physicist (even a theoretician) should be advised to keep away from it – this is a book for mathematicians.

He pointed out that there was a shorter road to the results that the great man had obtained and irreverently wrote that "Weyl combined much in his usual manner the superficial mathematical varnish with the poverty of physical ideas" when speaking about the most difficult questions of the quantum-relativistic theory. The verdict was harsh: "the book is hardly superior to other works on quantum mechanics where the physical material is concerned, with the exposition being probably unnecessarily involved", so therefore it was not worth being translated into Russian.

When reading about the early thirties, one may get an impression that the younger generation of physicists was overenthusiastic about theoretical issues; their critics even coined a special term to brand them with, "Talmudism" meaning a disregard for the phenomenological approach and the evergreen tree of physical life in favor of barren theoretical constructs based on the "first principles".[7] It should be said that Ioffe was among the critics – as an experimenter he seemed to find it hard to accept the changing style of theoretical physics.

Paradoxically enough Bronstein was free of an undue reverence for mathematics that was, after all, the language of theoretical physics. He wrote about Dirac's book:

"the simple and clear physical results are not spoiled with mathematical pedantry." Only a physicist could show little respect for Weyl's book, who was a prominent mathematician after all.

The rapid mathematization of theoretical physics that took place in the twentieth century did not close the gap between the two trades, and their world-views remained different. Weyl was a mathematician, though with the first unified field theory and the idea of gauge symmetry he got his place in the history of physics as well. To put things in a nutshell, one may say that while a physicist is seeking the unique truth and the unique mechanism of the Universe, the mathematician analyses the constructs he himself devises in an attempt to obtain all possible truths.[8]

Bronstein was not a simple "consumer" of mathematics; different from Landau he did not regard mathematics as an "labor implement". Always eager to learn more, he was prepared to wander into unknown mathematical fields. One of his friends told us that having bought Ingham's *Distribution of Prime Numbers* (1936) Bronstein on the same night was enthusiastically discussing it over the phone.

In the days when there was not a single systematic exposition of quantum mechanics in Russian, and it was considered to be extremely difficult if not irrational, Weyl's book was hardly suitable for the Russian reader. When declining the book Bronstein could not even imagine that the first Russian translation would appear 55 years later.

## 3.7. Cosmology in the Early Thirties

Bronstein's scientific interests embraced all of fundamental physics. He began his first year at the Leningrad FTI with quantum physics and finished it with relativity theory. Together with V. Frederiks he produced an encyclopedic entry about the theory of relativity while the *UFN* journal carried his vast survey of cosmology.

It was well-timed: consistent red shifts in the spectra of distant galaxies discovered by Hubble in 1929 became the first observational fact in cosmology. It turned relativistic cosmology, which had been a physical logic system into a genuine physical theory.

Bronstein felt at home in relativistic cosmology, the crossroad of astronomy, physics and mathematics. He cooperated with many astronomers and had been active in astrophysics long enough to deal with the unique material that could not be reproduced experimentally and differed greatly from what the physicists used to work with. The survey showed that his knowledge of physics and mathematics was fundamental and that he was able to present his material clearly. It was quite an event in the history of physics in Russia: to this day elderly physicists can remember the impact it produced.

As its title "Relativistic Cosmology Today" suggested, the survey drew a wide picture of the contemporary situation: nine out of twenty-nine works quoted by Bronstein were dated by 1930, the year the survey appeared.

The graphic and succinct introduction presented basic astronomical descriptions of the universe's stellar and galactic structure. Bronstein deemed it necessary to note that

> ... no matter how much the longsightedness of astronomical implements increases, an observer can never aspire to grasp the world as a whole. This may produce an impression that the cosmological problem is a stronghold that can never be taken by empirical science. Yet, where an observer despairs by his impotence, a physicist will undertake to solve a hopeless problem.

It was the author of General Relativity who first pointed at a physical approach to cosmology. Bronstein was fully aware that he was confronted with a special problem and that the physical object – "the world as a whole" – was unique. (Today we would use the term "universe", which is less precise.) For several decades cosmology remained outside physics because of its unique object, its narrow empirical foundation, its absence of geometrical limits and its infiniteness from the viewpoint of mathematical physics.

In his review, Bronstein made no attempt to camouflage the unique nature of the cosmological problem behind mathematical formulae: he was striving to reveal the mechanism of relativistic cosmology. He gave a concise description of Riemannian geometry to help his readers overcome the natural consternation caused by General Relativity's complexity and the paramount nature of the cosmological problem. He discussed three models of the universe that had been formulated by that time: static (cylinder) model of Einstein, de Sitter's model and Friedmann's–Lemaitre's non-static model.

We have already mentioned A. Friedmann. Bronstein came to the Main Geophysical Observatory in 1929 when the spirit of its former director was very much alive in the stories of Friedmann's former colleagues. Bronstein deemed it necessary to give "the late Russian mathematician who had introduced the non-static model of the universe back in 1922 and whose work had been practically forgotten" his due.[9]

In his survey, he discussed the three cosmological models together with the references to astronomical data obvious in his time. The result was a clear-cut definition of the "radius of the universe" that puzzled quite a few. He used both formulae and his common sense to demonstrate that "if the radius of the universe is very large, then the universe's cylinder shape (Einstein) affects but little the phenomena limited by small areas in the same way as the Earth's global shape practically does not influence what is going on in a room." Bronstein discussed only closed models (more favored in his time) though he also referred to Friedmann's work of 1924 that considered the case of a negative curvature, that is, an open infinite model.

He found convincing examples to explain the amazing properties of the relativistic geometries expressed in integrals and equations. Here is one of them, designed to clarify the concept of horizon: "In the world of de Sitter, letters sent to a point $R\pi/2$ away from the nearest post office will never arrive despite the post despatching them with the velocity of light."

By that time, relativistic cosmology, a more or less recent addition to the world of knowledge, had lived through several dramatic discussions. It was not Bronstein's intention to erase the delusions of his eminent colleagues. He was straightforward when he wrote that Weyl and Eddington were trying "by hook or by crook" to employ de Sitter's solution to explain the empirical domination of red shifts in the galactic spectra. He stated that the American variant of de Sitter's article carried an incorrect formula which was corrected in the Dutch variant.

His conclusion proved to be prophetic: "There is no doubt that many changes are in store for the cosmological theory. What is more important, it will have to extend its pinching time limits".

One cannot but wonder how a recent university graduate who gave much of his time to other fields of physics could produce this long and detailed survey. Indeed, in 1930 he was more than just another university graduate: his profound knowledge and wide-flung scientific interests made him a class in himself.

He had the situation in General Relativity at his fingertips: no wonder, therefore, that together with Frederiks he contributed an article on General Relativity to the "Technical encyclopedia".[10]

It was very typical of the time and the country it was published in: there was a general obsession with learning and accumulating scientific facts. Today we find the selection of subjects rather strange: relativity theory was discussed between Clay Processing and Heating (according to the Russian alphabet) in the volume that opened with Olive Tree and closed with Patent Law. All entries on physics were of a high standard.

Bronstein and Frederiks supplied all the principal physical sections (including cosmology). V. Fesenkov wrote about GR's astronomical verification while Timiryazev contributed a section on its "philosophical verification."

These sections give an idea of the status of relativity theory in Soviet science. Fesenkov, a prominent figure in Soviet astronomy, was very cautious: "Today GR cannot be verified through astronomical observations. It can be said, however, that not a single observable phenomenon contradicts it."

Timiryazev believed that "the basic propositions of relativity theory are incompatible with materialist dialectics." Though being much more reserved than in his other articles, he was out to demonstrate that the theory was "idealistic". By doing this he was opposing those who regarded the theory to be "a concrete form of the dialectical materialist teaching on space and time" (according to Timiryazev, they were B. Hessen, S. Semkovsky and O. Shmidt.)

It seems that credit for a clear exposition of the theory's physical side should go to Bronstein. Being educated in the pre-relativistic age, Frederiks was under the spell of both relativistic ideas and the methodological prejudices of those who inspired them. His 1921 survey [238] exhibited Poincare's conventionalism, Einstein's bias toward Mach's principle and Hilbert's axiomatism. He was too uncritical and steadfast in his adherence to Einstein's conviction that an absence of matter resulted in Euclidean geometry.

Bronstein was one year younger than the theory of relativity; therefore, it was much easier for him to work out its independent treatment. There are many places in the article that bear witness to Bronstein's independent and profound understanding.

One can easily picture his amazement when he opened a fresh issue of *Uspekhi fizicheskihk nauk* with his article prefaced and annotated by the editorial board. It takes no wisdom to guess that these philosophical commentaries and ridiculous alterations belonged to Hessen, the editor of *UFN*.

## 3.8. Ether and the Theory of Relativity

This commentary caused quite a stir in the Leningrad FTI; it triggered a scandal later known as "Hesseniad".

Boris Hessen (1893–1936), corresponding member of the USSR Academy of Sciences, Director of the Institute of Physics at Moscow University and Dean of the university's Physical Department, was also a prominent philosopher and historian of science. His paper "The Social and Economic Roots of Newton's Principia" [162] was favorably received at the 1931 London congress on the History of Science [175, 232].

One should mention that Hessen's own "social and economic roots" were no less remarkable. Through Reiser, Hessen's cousin, Bronstein learned that Hessen had been born into a wealthy family of a bank manager in the south of Russia. He became a social-democrat at an early age together with his close friends Tamm and Zavadovsky (both of whom were destined to go far in science). He did his bit of clandestine activity; after the 1917 revolution he confiscated the parental bank in the name of the Bolshevik party.

After Hessen graduated from the Institute of Red Professorate, which trained lecturers in the social sciences, he did a lot to turn the Physical Department of Moscow University into a modern research and training center. His efforts in this field were highly successful [257]. Besides he tried to fit the latest physical discoveries and theories into Marxism though he lacked the knowledge to assess adequately GR and quantum mechanics. The uncompromising young theoreticians unmercifully ridiculed him despite his heated defense of new physics.

To make things worse for Hessen, a volume of the *Greater Soviet Encyclopedia* with his entry on ether appeared in 1931 [161]. It was Bronstein who discovered it and shared his amusement with friends. Indeed, such statements as "the problem of ether is one of the greatest challenges in physics" or "it is one of the gravest methodological errors of GR to regard ether as an absolutely continuous matter" or "ether is an objective reality together with other material bodies" did nothing to boost his authority with the younger generation. These statements could have been regarded as gospel truth before 1905; in 1931 they were ridiculous. The fact that Hessen was one of the two editors-in-chief of the encyclopedia's division of physics added piquancy to the situation.

The young physicists despatched a telegram to Hessen addressed to the "Department of Exact Knowledge" of the GSE:

> *Having read your entry on ether started our enthusiastic studies of it. Looking forward to reading about phlogiston.*
> *Bronstein, Gamow*, Ivanenko, Izmailov, *Landau*, Chumbadze

This was a photo-telegram, a recent innovation that allowed one to send drawings together with the text. This particular telegram showed a dustbin with an old bottle labeled "phlogiston" together with a chamber pot labelled "ether".

Retaliation was a noisy and crowded meeting at the Leningrad FTI; the older generation was inclined to condemn the rude joke though not inclined to revive the idea of ether. Ioffe, who shared the post of the editor with Hessen, got separated from Leningrad – after all, Hessen was a philosophical dam that protected the newest physics against the flow of nonsense from Timiryazev and his cronies.

As a result, Bronstein and Landau were temporarily barred from lecturing at the Polytechnic [102]. Gamow told this story (inaccurately) in his autobiography [154].

Back in 1905, relativity theory did not do away with ether once and for all as one could imagine; in fact Hessen's article was not as absurd as it should have looked otherwise. Ether proved to be more viable than other fluids. Indeed, special relativity (SR) left no space for ether, which was completely replaced with the electromagnetic field. Minkowski geometry allowed no degrees of freedom for ether to be reintroduced.

General relativity changed the situation outright – the theory was richer by ten new values that could have been called gravitational potentials. Since in GR the gravitational field is coupled with geometry, there is every reason to believe that the new variables described the state of space-time whose universality allowed ether to survive in a sense.

In 1920 many in the physical community were relieved to hear Einstein himself discussing a relationship between ether and GR. Obviously, the older generation, who had entered science before GR, found it hard to part with the idea of ether. They were in good company: Lorentz and Poincare were reluctant to abandon ether despite their own contributions into special relativity. It was not quite by chance that Einstein deemed it necessary to "make peace with ether" in Leiden, Lorentz's city. There was more than mutual empathy between the two physicists behind this – the well-educated Leiden public gave a chance to analyze the General Relativity's fundamental ideas.

In actual fact, this changed nothing in the theory; Einstein just made it easier for the older generation to regard space-time as a dynamic rather than a static system. They would rather have accepted a character named Ether (less a material than a host) in the physical play staged by Einstein than to reconcile with the stage playing an active part in it. Lorentz was one of those who had done a lot to make ether invisible and immaterial, and special relativity just put the finishing touches on his efforts.

Einstein himself did not need ether as a physical idea, although his first scientific paper dealt precisely with ether [304]. In his mind, ether had been practically substituted with the idea of space-time. The substitution was not complete: since the very beginning Einstein was nursing an idea of generalizing it by incorporating quantum ideas to create a unified theory. Naturally, this prevented space-time from becoming absolute.

Even younger physicists failed to grasp the true meaning of the Leiden speech; Sergei Vavilov wrote at that time: "It is significant that the man who had banned the idea of world ether removed the ban himself. For fifteen years his hypnotic influence braked the natural development of the hypothesis that was undoubtedly valuable for physics" [122]. This was said by the man who in 1928 wrote in his book *Experimental Foundations of Relativity*: "The empty space of Democritus and Euclid and ether that was hard to grasp was removed in favor of a complicated yet physically tangible space-time of Einstein" [123, p. 13]. In his time this book was another step toward granting GR its place in the sun.

There were physicists had accepted the substitution of space-time for ether as a fact and saw beyond the mere words of the Leiden speech. Hessen was not among them. In his book [160] and the encyclopedic entry, he demonstrated that his knowledge of SR was quite adequate but his ideas of GR rather superfluous – his loyalty to ether was rooted in pre-relativistic physics. This was what invited the young physicists' ridicule.

There are no grounds to believe that Bronstein was a more ardent relativist than Einstein and a more violent opponent of the idea of ether. In 1929 article "Ether and Its Role in Old and New Physics", Bronstein looked back at the concept's history and drew the rather unexpected conclusion that theoretical physics could not afford to abandon it altogether.

This conclusion was not quite unexpected: back in 1924 and 1930, Einstein made similar forecasts [309,310]. One can say that they proved to be in vain if ether is to be taken literally as the idea burdened with past associations. Ether as an ever-present universal physical environment has survived under different names. Today it is very visible in physics. In the twenties it assumed the names of space-time and unified field. After a long interval it emerged into the foreground once more. Vacuum is one such name; it can be heated and can take part in phase transitions. The old supporters of ether would have been delighted and puzzled at the same time. It would have never entered their heads that it could produce particles. The key problem of the new theory of "ether" was a unification of GR and quantum theory. Bronstein came up with it back in 1935. We shall supply more details in Chapter 5.

## 3.9. Styles and Generations

The relationships between "fathers and sons" determined the general atmosphere in Soviet theoretical physics.

Genrikh Neighauz, an outstanding musician and teacher, used to say that while it was outside one's power to create talents one could, at least, tend the soil on which they grew. The same can be said about any other field of human endeavor. It was under "Papa Ioffe's" patronage that young physicists grew into prominent researchers of international renown. Numerous reminiscences and letters testify that Yakov Frenkel, who headed the theoretical department, was the director's match in this respect [139, 284]. He figured as Kind in Berestetsky's essay that we have already mentioned above.

Bronstein received his share of this kindness as well. In a letter dated November 1930 Frenkel wrote: "I am extremely lucky with my associates ... . The Abbot is the most talented of them." He used the Odessa Congress to make arrangements for Bronstein to work with Sommerfeld; during his stay in the United States (1930–1931) he did his best to procure a Rockefeller grant for him [284]. (Frenkel, Krutkov, Fock, Skobeltsyn, Gamow, Landau and Shubnikov had already used this opportunity to work abroad in the best scientific centers.)

The country's scientific potential was swelling as more and more young people joined the academic community. When summing up his institute's first fifteen years, Ioffe wrote in 1933: "We rely on young people in our creative work – this is our key principle. Some people ironically dubbed us a kindergarten" [194].

The young and talented researchers showed little respect for the "old people" of forty or so; as often happens in the family they never showed enough appreciation for Papa Ioffe or, rather, they believed that he was abusing his parental rights. He valiantly fought every time his "children" were attacked by ignorant and zealous ideologists; he never tired of creating an atmosphere where talent would flourish.

Here are some typical episodes borrowed from V. Saveliev's reminiscences and a series of "on-site reports" sent by Nina Kanegisser to her sister who had moved to Zurich upon her marriage.

Saveliev met Bronstein in 1931–1937 as a student and a post-graduate student. He wrote [261]:

"M.P. and Landau were worlds apart. Mild and witty, Bronstein had little in common with unkind and sarcastic Landau. He never failed students at exams: pleasantly surprised with at least some knowledge in students, he would give everybody excellent marks. I suspect that my own excellent mark in statistical physics I got from him worth very little.

"I also remember some of his 'acting': it was usual to accompany all official celebrations with some of amateur theatricals. I see distinctly in my mind's eye a table on the stage in the large assembly hall; Frenkel was sitting at it with a voltmeter and a voltameter in front of him. He says to Bronstein: 'They accuse you and all other theoreticians of our department of idealism and total disregard for practice. Can you shame your critics by stating the purposes of these devices?'

"Without even turning his head, Bronstein answers: 'This is all very simple – voltmeter measures voltage while voltameter measures amperes as well.'

"'I should say that with your logic you would shame all your critics; could you please show me which is which?"

"With a sweeping gesture designed to conceal his 'ignorance', Bronstein points to both devices. Frenkel goes on with the test and puts one of the devices behind his back. Here again, Bronstein is ready with an answer: 'The device did not disappear when you put it behind your back, but it has ceased to be an object of physical research'. The delighted audience appreciates the verbatim quotation from Ph. Frank, a famous positivist philosopher and physicist who replaced Einstein at the department of theoretical physics in Prague.

"On another occasion they showed us drawings by Yuzefovich projected onto a screen. In one of them Frenkel in sports attire was taking three puppies with the heads of Bronstein, Landau and Ivanenko on leashes. The inscription said: 'Lately, Frenkel and his pack go astray far too often'".

This drawing was certainly made in late 1931 or early 1932, when both Landau and Ivanenko were still at the Leningrad FTI. In August 1932 Landau moved to Kharkov; some time later his friendship with Ivanenko ended.

Another friendship proved to be more stable; Jonny, Dau and the Abbot (as they were known at the institute) were invariable participants, if not initiators, of many practical jokes and scandals. In fact, their elders preferred to call them *Drei Spitzbuben*.

Though physics was their main concern; they wanted to organize training and research in a manner that would rocket science in the Soviet Union onto the world level. Some of these plans were later embodied in theoretical courses and a textbook of theoretical physics. They wanted to set up an institute that would expand studies of theoretical physics; they felt that their generation and their "theoretics" were poorly represented at the Academy of Sciences. There was a tacit agreement that Gamow of the younger generation had a greater chance to be elected. After all, his theory of alpha-decay (1928) was recognized world wide.

Here are some extracts from Nina Kanegisser's letters to her sister Zhenya [246]. They give ample ideas about the Jazz-band and the events with which it was involved. It is very hard to tell truth from invention in them; the author, being a biologist, might find it hard to sort out her information. One feels, however, that on the whole the descriptions are true-to-life.

> The hottest news of the season is an attempt to make Jonny an academician. Dau and the Abbot decided between themselves that Jonny should become an academician to promote theoretics. Luckily at that point Dau fell ill, and it was the Abbot's lot to put this crazy idea into life. He went so far as to consult Abrau [Ioffe] about their plan. Abrau mildly stated that the idea was at best ridiculous. You can imagine that the Abbot was enraged; being fully aware that Jonny stood no chance, they were mulish about it. They were planning a campaign in the press to turn the public against the grand old men who were unwilling to accept young talent into their company. They even wrote to Bohr to ask him for recommendations (to make them public).

This was followed by more contributions.

> There is a lot of commotion around the comedy called "Jonny the Academician". The ripples reached foreign countries. Kapitza (to whom they had written to enlist his support and that of Rutherford) replied from Britain: "I fully agree that we need to rejuvenate the Academy and that Jonny is a suitable animal for this experiment. Being no Dr. Voronov, I resolutely refuse to interfere with the affairs of others.[11] Isn't this funny? To sum up – a quarrel with Yasha [Frenkel], who was seeking the same post himself, and strained relations with Abrau. Dau and the Abbot are waging the campaign, while Jonny modestly repeats that he can be satisfied with a corresponding member. Dau and the Abbot are toiling while Jonny is counting his future salary. It seems that he alone cherishes some hopes and is sincerely disappointed when told that nothing will come of it.

[Nevertheless in 1932 Jonny was elected corresponding member.]

> There is a new putsch in the theoretical sector. Jonny is Director with Dau, Ambartsumyan and the Abbot are members together with a young Georgian post-graduate student who is overenthusiastic over them. The triumvirate does not want either Frenkel or Dymus. Ambartsumyan seems to be less strict; a scandal followed. Dymus is mad at Jonny and Jonny at Dymus because he did not vote for his candidature for the academician. I should say that he was supported only by the Abbot, Dau and the mad Georgian.
>
> There are more amusing developments on the Triumvirate-Abrau front. By the way, Yasha calls them Hamow, Ham and Chameleon. Not bad, is it? At a seminar on materialist dialectics, somebody accused Yasha, who was busy demonstrating his materialism, that his pupils were not materialistic enough. Poor Yasha retorted: "You can take these pupils. These serpents are pecking me to death". The pecking serpent Abbot has finally quarreled with Abrau because of a radio speech about physics. One fine evening at home Abrau switched on the radio only to hear the Abbot's creaking

voice denouncing the law of energy conservation by quoting from Bohr, Dau and Rudi [Peierls; for more details see Ch. 4]. After the talk came an announcement: "You listened to such and such scientific associate with Ioffe's Institute". In the morning, enraged and frustrated, Abrau summoned the Abbot to his office; a tempestuous scene followed, during which the serpent insisted that Abrau had exceeded his authority (it seems that he was right in that); he was quite impudent and handed in his resignation. The storm abated after a while: the Abbot seems to be tired and sad.

One may think that Frenkel passed too harsh a judgment on his young colleagues, in his nicknames. It was not a judgment – he just retaliated in the same witty manner as he was attacked by the Triumvirate. This was the time when ready wits were considered an academic weapon and when titles and posts were not enough to earn authority; after all, one could be a student, a post-graduate student and a head of the department of physics in a "communist college", where managers of all ranks upgraded their education at one and the same time.

There is a grain of truth in every joke. While the hint is quite clear in the first nickname – Hamow, the second was prompted by Landau's famous tactlessness. As for the "Chameleon" that was related to Bronstein, it reflected his tact toward those who thought differently from him. Being a completely independent and critically minded person he was tolerant – the trait that Landau regarded as his fault.

It seems that Bronstein did not much like Frenkel's scientific style: he was too proficient with ideas and images and not strict enough in their formalization. He thought in models rather than exact formulas. This was sufficient for him – his interest cooled down and he turned to another problem. In fact, Landau's style of thinking was much closer to Bronstein's who, however, did not share Landau's conviction that only one style was acceptable in science.[12] In any case Bronstein's attitude to Frenkel was never affected by the difference in their styles.

It seems that this tolerance prompted the nickname that could have been rather offensive had not all been aware of Bronstein's firmness over the matters of principle.

Age alone was not enough to explain the differences in styles – it goes back to the personality structure. The seemingly simple methods of explanation that Frenkel employed looked rather vague not only to his young colleagues. Here is what Ehrenfest wrote to his friend Ioffe in 1924: "Frenkel's style of thinking is so different from mine that a fruitful cooperation is hardly possible. He is infinitely more interested in results than explanations – so far, at least, we have failed to find a common language. It seems that Pauli, who is a fast and precise thinker can cope with him" [311, p. 178].

Here the word "explanation" should be explained. More often than not new general principles in physics are not based on firm ground; a solid foundation of a minimal number of non-contradictory axioms is built much later. Ehrenfest's critical mind and sharp perception demanded an "explanation", that is, fundamental axioms and well-substantiated patterns – for him this was indispensable. In his turn, Frenkel believed that to prove the correctness of a new theory one should resolve it with the help of old problems and formulate new ones. For him this was much more important than Ehrenfest's explanation.

Ioffe was obviously on Frenkel's side; the theoretics of his younger colleagues was too abstract for him. By contrast, Ehrenfest, who was closely watching the younger generation of physicists of the Soviet Union, openly supported them. This was especially important for Landau, who in 1932 had to go to Kharkov because of strained relations with his older colleagues in Leningrad. Ehrenfest came to know him better in the winter of 1932–33. He wrote to Ioffe on January 6, 1933 [311, p. 262]:

> There is no doubt that experimental physics needs theorists like Frenkel, Zernike or Ornstein .... It is doubtful, however, that they are able to contribute, directly or through their pupils, to theoretical physics .... On the other hand, Landau (despite his detestable misbehavior) is the type of a physicist-theoretician *absolutely indispensable* for any country. One should admit that something of a talmudist is present in his style of thinking (I can say the same about Einstein and myself). In any case, it is more prominent in his talks than in his thinking. A couple of heated discussions about his unfounded paradoxical judgments convinced me that he is a *clear* and *graphic* thinker – especially in classical physics.

Young theoreticians found it hard to accept the style they did not approve of – "live and let live" feeling comes with age. A. Migdal, who in 1936 started as Bronstein's post-graduate student, wrote:

"Yakov Frenkel took over my post-graduate training after Bronstein. Our styles were worlds apart, besides there was another thing that interfered with our cooperation. I regret to say that I and my young colleagues failed to appreciate this original and bold thinker; all of us were captivated by Landau, who insisted on quantitative solutions and looked down on the qualitative ideas with which Frenkel was brimming. Much later I was amazed to discover that Frenkel had influenced me immensely to the extent that my style became much closer to his. And this despite the fact that I was never close to him though I admired his splendid personality" [238, p. 23].

## 3.10. The Physics of Semiconductors and Nuclear Physics

One should not imagine that Bronstein was totally wrapped up in fundamental physics: a true researcher is easily ignited by an interesting problem. Einstein's mind was easily stirred by the behavior of tea-leaves in a cup, the shape of a river-bed, or better aircraft design [288].

Bronstein also suggested an electromagnetic method of determining aircraft speed that was approved by Kurchatov; it seems that writing for him was akin to innovation – the task was how to explain complex things in a simple and elegant way.

The development of physics and his own duties at the institute posed a host of research problems; in his CV of June 23, 1935, he wrote: "I have written a number of works on the theory of electrical semiconductors, cosmology and other problems; at the same time I have done some teaching at the university and the Physicotechnical Institute, where in the 1934–35 academic year I read a course on the theory of nucleus for younger colleagues. I have also written some surveys and popular science books, including *The Structure of Matter*, 1935" [103].

One cannot but be impressed by the wide range of his scientific interests – after all, semiconductors and cosmology have very little in common. One will be even more impressed with the two dozen scientific papers, twenty-odd popular science articles (including two books) and a children's book, *The Solar Matter*, that he penned during his first five years at the institute. Actually his range of lectures for students and post-graduate students was much wider than he stated – they covered electrodynamics, quantum mechanics, statistical physics, General Relativity, etc.

His work on the theory of semiconductors [193] was well received by the academic community: in 1934, when academic ranks were reintroduced, everybody at the institute (Frenkel and the Learned Council in the first place) felt that it should earn Bronstein a doctor's degree (he himself decided differently).

The institute launched the studies in the early thirties and Bronstein immediately became part of the project [12, 17]; his review [13], intended mainly for the experimenters, was instrumental in boosting the studies. With his usual skill he gave a clear and concise description of Wilson's main results and offered a quantitative theory of conductivity of semiconductors and of the thermoelectric, galvanomagnetic and thermomagnetic phenomena in them. Bronstein stood at the very source of the wide flow of the physics of semiconductors; no wonder, therefore, that his studies were drowned in later research.

Bronstein did not write much on nuclear physics – his enlightenment efforts proved to be more important. He helped assimilate the landslide of new knowledge in this field. When it was young, nuclear physics was ill-suited for theoreticians of Bronstein's type; it was the natural prey for pragmatically minded theoreticians who didn't care about the integral picture of the world. Before 1932 everybody agreed that without a relativistic quantum theory there was little hope of probing nuclear phenomena (the theory of electrons inside a nucleus). This conviction was shattered by the discovery of the neutron and the hypothesis about the neutrino.

Late in 1932 Bronstein was included in a group considering a problem of nuclear physics that was officially called "the second central problem of the institute's research range" [132]. (The first was the physics of semiconductors.) Ioffe was appointed the head of the group, Kurchatov was his deputy, while Ivanenko was responsible for the seminar on nuclear physics.

Bronstein took an active part both in the nuclear studies and in the seminar's activity. He was a theoretician in the department of nuclear physics and read lectures on the theory of the nucleus for the institute's staff; he was a judge of Lev Artsimovich's thesis [286] and wrote frequently about nuclear physics in popular journals, books and the encyclopedia.

The First all-union conference on nuclear physics proved to be a signal event in the Soviet nuclear studies. The decision about it was made in late 1932; Kurchatov was to head the organizing committee [287]. The conference took place on September 24–30, 1933. The year 1932 proved to be a great year for nuclear physics when the neutron and positron were discovered. These experimental discoveries posed some urgent questions about the nature of the positron, the nature of cosmic rays (in which positrons were discovered), the structure of the nucleus, the physics of beta-decay and the anomalous scattering of gamma-rays, etc. The conference was intended to discuss these sensational new ideas and the theoretical problems they posed.

The situation in physics was dramatic: the discovery of the positron was a triumph of Dirac's theory and put him into the forefront. However, the infinite (and unnoticeable) density of the charge and energy of electrons at negative levels suggested by his theory put the theoreticians into a quandary. Fock was especially eloquent about this at the conference. Still, not everybody accepted the neutron-proton model of the nucleus and the neutron's elementary nature. What was more, Dirac, Joliot and some other physicists voiced their strong doubts about the elementary nature of the proton. There were several reasons for this: indications that the mass of the proton was larger than the mass of the neutron; some observational data pointing to positron radioactivity and, last but not least, the proton's measured magnetic moment, which turned out to be much larger than could be expected from Dirac's equation.

The newly acquired material that was burning in everybody's mind was being assimilated when the conference published its proceedings. Bronstein was the conference secretary. It was he who prepared various surveys [19, 77]; he finished them days after the conference folded; the emotions that bubbled in them made even those who had been absent from the conference its participants.[13] The survey in *Sorena* [77] even carried small drawings of likeness of all those present at the conference done by N. Mamontov. Bronstein made public the experimental and theoretical results that had not appeared in the press; he also discussed a work by Dirac and Peierls that had not quite finished by that time. It dealt with the "deformed, or polarized, distribution of electrons with negative energy". Modern terms describe this as the polarization of vacuum.[14]

Bronstein himself spoke on September 27 at the conference on cosmological problems together with Dirac (on the positron theory), G. Bek (on the theory of continuous beta-spectra), V. Weisskopf (on new theory of Bohr and Rosenfeld).

At first glance Bronstein seemed to wander far from the conference. The preliminary program drawn in December 1932 contained a section on the theory of nuclear structure and the problems of relativistic quantum mechanics; Bronstein was supposed to speak on the subject [132]. In 1933 Bronstein published three works on cosmology that give a good idea of his conference paper; they were directly related to relativistic quantum theory in the same way as all the other reports read on the same day. In his article [16], Bronstein made an attempt to wed cosmology and Bohr's hypothesis on the violation of the law of energy conservation in quantum-relativistic physics (in the pre-neutrino age, beta-decay belonged to it). Another article [22], written together with Landau, dealt with the same subject; still another article [21] discussed the position of the cosmological problem in the structure of a complete physical theory. We shall discuss more details below. It is very hard to stick to chronological order when writing about such a versatile man.

Nuclear physics was next discussed between May 20 and 22, 1934, at a conference on theoretical physics in Kharkov. Bohr came together with Rosenfeld. (On May 20 the *Kharkovski rabochii* newspaper carried a photograph showing Landau, Bohr, Rosenfeld and Bronstein together. Later Bronstein published a detailed report in *Uspekhi fizicheskikh nauk* [26]. In his report he tried to tie nuclear physics to astrophysics by probing the origins of cosmic rays and explosions of supernova. Nothing came of it; both scientific fields were too young at that stage.

The second nuclear conference was intended to take place in September 1935 but was actually convened much later, in 1937. There is a letter from Bronstein to Dirac dated April 21, 1935, inviting him to come. He also informed Dirac that the second edition of his book (translated into Russian by Bronstein and Ivanenko) "will appear very soon, in two or three months". The book appeared in 1937. On April 11, 1937, Bronstein wrote to Fock:

"Today I signed Dirac's book to be sent to press. Unfortunately, this time I lost the battle I was waging for this book with the scoundrels in the publishing house. First, they insisted that Dymus' name to be removed from the front page; to balance things I removed my own name as well, though preserved it as the name of the editor. I have all the right to do this since I corrected what Dymus had done. Second, they prefaced it with an indecent piece, explaining that Dirac is a villain" [90].[15]

Fate decreed it that the last of Bronstein's articles was to be on nuclear physics as well. He calculated the influence of the neutron's magnetic moment on the interaction with matter where it was moving. He did these calculations at Kurchatov's request; they were connected with the experiments he planned at the Teachers' Training College where Kurchatov headed a department. It seems that the article was not without the pedagogical aim of teaching the experimenters to handle the general methods of quantum mechanics to tackle specific problems. Kurchatov's associates said that Bronstein had often spoken on modern physics at the college.

In April 1937 Bronstein wrote to Fock that he was working on a detailed article on the anomalous scattering of electrons with the nuclei (preliminary note [32]), he connected the anomality with beta-interaction. It seems that Mandelstam and

Vavilov had this work in mind when they wrote in 1938: "in a number of his works on the physics of the atomic nucleus Bronstein has demonstrated in which particular phenomena the exchange nature of nuclear forces was most prominent".

His scientific activity and his teaching efforts contributed to the speedy development of nuclear physics in the Soviet Union when the situation was ripe.

# Chapter 4
# Hard Times for the Laws of Conservation and for Theoreticians

Anybody wishing to get a clear idea of Bronstein's position would be amazed with his popular article "Is Energy Conserved?" written in 1935. Today the law of conservation of energy is part of the foundation of the current picture of the world. One cannot but be astonished at the arguments and heated agitation against the law that stretched far beyond physics. To prove his point, Bronstein wrote, in particular, that the law was very much "like tidy accounts with all pennies counted – a sight that gladdens the bourgeois heart". He even suggested that the *perpetuum mobile* that would work on the non-conservation of energy in the quantum-relativistic field was a potential starting point for the technology of future communism.

An embarrassing statement indeed. What prompted the article? Let's have a closer look at the events that led to it, the scientific context of the thirties and Bronstein's picture of the world.

Today they prefer to ignore the fact that in the thirties the law of conservation was under attack. Many outstanding physicists – Landau, Gamow, Peierls and Dirac of the younger generation and Ehrenfest of the older generation – cast doubt on it; it was Niels Bohr, one of the greatest physicists of the twentieth century, who suggested the hypothesis of non-conservation.

In the twenties and thirties there were three attempts to topple the law of conservation of energy; Bohr was directly involved in the first two of them; he participated in the third one indirectly.

There is a vast body of literature dealing with Bohr's heritage that laid the foundation of contemporary science; his role in twentieth-century physics cannot be overestimated. However, *errare humanum est*. This chapter deals with one of Bohr's errors, the greatest blunder in his scientific career: the hypothesis that the law of conservation did not apply to subatomic physics. A superficial observer might think that the hypothesis was more than erroneous – it was totally unfounded; a deeper probe into scientific development might reveal a different picture. After all, errors can speak volumes about the state of science and a scientist's methodology. It was not on the spur of the moment that the great physicist formulated this hypothesis: he pondered it for more than a decade – between 1922 and 1936.

Why did a seemingly still-born hypothesis prove to be viable?

## 4.1. Three Attempts to Topple the Law of Conservation of Energy

Bohr first formulated his idea of the limited applicability of the law of conservation in subatomic physics in an article written in November 1922 and published in 1923 [113]. He was led to it by his contemplation of the gap between the wave description of light and the idea of light quanta first introduced by Einstein in 1905 and later called photons. At that time, Bohr was mainly operating under the principle of correspondence and failed to embrace both wave theory and light quanta. Therefore, he discarded the light quanta idea. Since Einstein's heuristic idea of light as a stream of quanta that successfully explained the photoeffect was based on the law of conservation, it was logical for Bohr to doubt the law of conservation. Some physicists shared his doubts [202, p. 133] but he alone proved bold enough to throw his authority publicly against the law.

Bohr was perceptive enough to see clearly the gap between the quantum discrete and the classical continuous descriptions and believed that the law of conservation was not too high a price for a bridge across the gap [241, p. 290]. His experience of formulating the theory of the atom taught him that sometimes you could not reach far by little steps. A physicist of his stature found it hard to reconcile the absence of what Einstein called the inner perfection of the physical picture; he was less impressed by the justifications of the idea of the light quanta achieved by 1922.

In 1923 the newly discovered Compton effect and its photon explanation based on the law of conservation propped up these justifications but failed to bridge the gap. Bohr didn't ease his pressure on the light quanta idea: the very next year he, together with Kramers and Slater, suggested a new approach to the Compton effect that excluded light quanta and limited the application of the law of conservation to its statistical meaning [120]. However, in the next year experiments (Compton–Simon and Bothe–Geiger) unequivocally pointed to the photon description.

The first attack on the law of conservation was repulsed. For Bohr it did end in 1927 with a desirable theoretical bridge of the uncertainty and complementarity principles that followed from a consistent apparatus of quantum mechanics; this bridge closed the glaring gap between the corpuscular and the wave descriptions of light and matter.

The second attack was launched by nuclear physics; unlike the first one, it was triggered by experimental deficiency and ended with a theory (very much fused with experiment). Let's look at how the events unfolded.

It all started in 1927 with the Ellis–Wooster experiments who determined that the electrons emitted through beta-decay were distributed along the continuous range of energies. Though the initial and final states of nucleus have definite energies, their difference is larger than the average energy of beta-electrons. It was established that beta-decay was not accompanied by the gamma-radiation that could reestablish the energy balance at every individual act of beta-decay. This led Bohr to the idea that the law of conservation might be violated in nuclear physics. A handwritten copy of a note he sent to Pauli in July 1929 and the letters that followed are the earliest evidence of Bohr's doubts [247, p. 4][1]. It was in October 1931 that he first publicly voiced his hypothesis [116].

Pauli saw little foundation in what Bohr wrote to him and responded with a hypothesis of his own. In December 1930 he sent a letter to a conference of "radioactive ladies and gentlemen" that gathered in Tübingen: "I have come upon a desperate way out regarding the 'wrong' statistics of the $N$- and the $Li^6$-nuclei, as well as to the continuous beta-spectrum, in order to save the theorem of statistics and the energy law" [252, p. 390].[2] He further continued: "There could exist in the nucleus electrically neutral particles, which I shall call neutrons, which have spin 1/2 and satisfy the exclusion principle .... The continuous beta-spectrum would then become understandable from the assumption that in beta-decay a neutron is emitted along with the electron, in such a way that the sum of the energies of the neutron and the electron is constant". This new particle would have averted also the "nitrogen catastrophe". Very soon, however, Pauli realized that the neutral particle alone could not be applied to both ailments. In June 1931 he publicly declared (in an oral statement) that the law of conservation could be saved with the help of "very penetrating neutral particles" that accompanied beta-decay [252, p. 393].

The two opposing hypotheses met at the nuclear congress in Rome in October 1931. There Fermi (who gave the new particle its name and formulated its theory) sided with Pauli. This was powerful support, but the rest of the congress was more inclined to side with Bohr. His idea was first published in the *Proceedings of the Rome Congress*. In his review, Bronstein wrote:

> According to Bohr, whose ideas are now generally accepted by theoreticians, the laws of conservation of energy and angular momentum (one of the most typical traits of the contemporary physical theory) are not relevant in the relativistic theory of quanta [68].

Pauli waited for the Solvay Conference of 1933 (October) to make his neutrino hypothesis public. There was also a communication about the clearly defined upper limit of the beta-spectrum that was in complete accord with the law of conservation; two new elementary particles – the neutron and the positron – had been experimentally discovered. The number of physicists who doubted the law of conservation shrank after the conference and, especially, after Fermi's theory of beta-decay that he formulated late in 1933. In 1936, after the short but dramatic crisis caused by Shankland's experiments there were no longer doubts.

The experiments staged by R. Shankland to study the Compton scattering in high energies seemed to prove that something was wrong with photon theory and the laws of conservation. The excitement they caused and the flare-up of discussions of whether the laws of conservation could be applied in the subatomic world can be explained solely by the childish faith that the third time is lucky. Very soon, however, Shankland's experiments were disproved and forgotten; all doubts about the laws of conservation dissipated.

Bohr was destined to put an end to the attempts to topple the law by his commentaries on the experiments that disproved Shankland's results:

"It may be remarked that the grounds for serious doubts as regards the strict validity of the conservation laws in the problem of the emission of beta-rays from atomic nuclei are now largely removed" [119].

The words "serious" and "largely" speak volumes; it was painful to sever the ties with the hypothesis of non-conservation. Much later, in 1957, Pauli was puzzled to note that

> it was only in 1936 when Fermi had successfully elaborated his theory that Bohr finally admitted that the law of conservation of energy was valid for beta-decay and recognized the neutrino [252, p. 394].

Let's take a closer look at these events and the participants.

## 4.2. The Hypothesis of Non-conservation and the Arguments of Its Supporters

### *4.2.1. Waiting for a Relativistic Theory of Quanta*

Bohr had found little support for his first doubts on the laws of conservation, rooted in his antipathy for light quanta. Even his associate, Slater, whom Bohr invited in 1924 to elaborate the non-conservation of energy was not inclined to doubt [202, p. 138]. At the same time, many physicists of all generations cast doubt on the idea of light quanta. In 1927, when discussing quantizing electromagnetic radiation, Landau said: "The light quanta were not strictly necessary; they were introduced arbitrarily" [213, p. 21]. Bronstein, on the other hand, supported the photon idea, as his earliest works testify.

It seems that the quantum paradox, as the problem of mating the discrete and continuous systems of description was called at the time, was a source of inspiration for those theoreticians who found themselves riding the crest of the wave. The very complexity held promise for a solution much more spectacular than the solution of the ether paradox by relativity theory. At the same time, the majority of theoreticians saw little sense in discarding the laws of conservation while a substitute was conspicuously absent.

In 1929, when Bohr revived his idea, the situation was totally different. The laws of conservation had failed to account for an experimental fact of the continuing beta-spectrum. What was even more important, the theory itself seemed to approve of radically changed behavior in the nuclear world. Before 1932 when the neutron was discovered there were no doubts that electrons were part of the nucleus – beta-radiation was the best evidence of this. The uncertainty principle formulated in 1927 demonstrated that a non-relativistic theory (that is, quantum mechanics) could not be applied to intra-nucleus electrons: by placing the size of a nucleus and the electron mass into the relation $\Delta x \Delta p \sim \hbar$, one obtains the relativistic velocities for electrons that needs relativistic theory.

One should also take into account the general situation in fundamental physics at the start of the thirties. It was the time of expectation of a relativistic theory of quanta that would equally employ the two world constants $c$ and $\hbar$. Dirac's equation for the electron (1928) was commonly accepted as an outstanding result although somewhat deficient in its want of negative states. Besides, the community of physicists expected much more of a genuine $c\hbar$ theory than Dirac's equation was able to produce. A synthesis of the relativistic and quantum ideas in a $c\hbar$ theory was expected as the final important event in theoretical physics; it was commonly believed that this theory would explain the numerical value of the fine structure constant $\alpha = e^2/c\hbar$ and, by the same token, the atomism of the charge [81, p. 205].

There were few people at that time who realized that the $c\hbar$ theory would inevitably bring a $cG\hbar$ theory in its wake and a cosmology based on it [21, 250]; the majority took the weakness of gravitational interaction and its negligibility for atomic physics as an excuse to ignore $G$.

In the late twenties the physicists had failed to adjust to the radical changes introduced by quantum mechanics but were sure that the coming $c\hbar$ theory would introduce even deeper changes [252, p. 72]. This assurance was prompted by several factors.

First, the idea of a unified field theory had not yet died, though Einstein's ideal of such a theory evoked little enthusiasm. On the whole a unification of relativism, quanta, gravitation and electromagnetism seemed to be accessible in the nearest future – this gave theoreticians their second wind.

Second, theoretical radicalism was spurred on by the difficulties (infinities of the field theory, for one) that could not be removed by the available means. The limitations of the conceptual apparatus resulted from both relativism and quantism was another, more important, source of radicalism in theoretical physics (individual uncertainties, lack of meaning of the "field in a point" idea, etc [158]). This radicalism was also fed by what seemed to be fundamental defects in Dirac's theory, the first ever quantum-relativistic theory. In 1932, when the positron was discovered, these defects turned out to be triumphs.

The significant breakthroughs of quantum mechanics convinced them that physics was moving in the right direction; the circumstances listed above showed that the desired aim was far ahead. At the start of the thirties theoreticians were propelled by the revolutionary impact of the age of relativity theory and quantum mechanics. They had adjusted to the tempo of the changes of the previous decades; this explains why the radical idea of quantized space-time was readily accepted and why there was no sharp response to the radical hypothesis of non-conservation.

### *4.2.2. The Neutrino Alternative*

These revolutionary sentiments were mostly responsible for the general negative attitude toward Pauli's neutrino hypothesis. It seemed too simple to be true.

It is easy to understand why it lacked charm in the early thirties. There was a general conviction that matter consisted of only two elementary particles – the electron and the proton – since their existence had been proved by a huge body of experimental facts. Both of them were electrically charged. The photon was not regarded as being on a par with the material particles. It was too recently discovered, and traditionally light was opposed to matter and mainly characterized the interaction.

It seemed that the particle that had no charge and practically no mass and, therefore, could not be traced, was nothing more than a futile attempt to save the old law singled out for replacement. This explains why for three years Pauli refrained from making his hypothesis public and restrained himself to oral discussions. In the autumn of 1933, on the very eve of a radical turn of opinion in the physical community, Bronstein wrote [77]: "Until quite recently it was considered bad manners to accept the neutrino; therefore all the physicists didn't waver in acceptance of Bohr's alternative", that is, the non-conservation hypothesis.

Only someone who was able to take a broader view of science's past and future could accept a changeable number of elements of which matter was composed. Bronstein wrote in 1930:

> The world proved to be simpler than the ancient Greeks assumed it to be. They believed that all natural bodies were made up of four elements, earth, water, air and fire. Today, there is a common opinion that protons and electrons are the final elements that form all material bodies. Will this conviction last? [63, p. 58].

Back in 1930 there were few physicists who would believe that it would last for so short time.

All the general methodological patterns in physics had not included the neutrino until two new particles (one of them electrically neutral) were experimentally discovered in 1932. In fact, before 1932 the neutrino could have been surmised only through theoretical empiricism – the nitrogen catastrophe, the upper limit of the beta-spectrum, etc. In fact, the law of conservation was not considered a theoretical priority: despite its significance and philosophical support the law of conservation was nothing more than a consequence of the developed dynamic theory as an integral of the motion equations.

The year 1932 proved to be the "the year of miracles" for nuclear physics: the discovery of the neutron did away with the electrons inside a nucleus and propped up the idea that the beta-electrons had been born. This sapped the central pillar of the hypothesis of non-conservation – the complete $c\hbar$ theory and nuclear theory turned out to be independent of each other to a great extent and non-relativistic quantum mechanics could be applied to a vast field of nuclear physics. Still, the draft program of the Leningrad Nuclear Conference (December 1932) carried "the theory of the structure of the nucleus and problems of relativistic quantum mechanics". At the conference held in September 1933 the neutrino proved to be peripheral – the report written by Ivanenko, one of the most active of the Soviet participants, did not even use the word [188].

Changes came late in 1933. In October 1933 the new experimental data about the upper limit of the beta-spectrum were made public at the Solvay Conference, so the neutrino hypothesis invited more attention. Pauli was prepared to publish his considerations while Bohr was more cautious in formulating his position. At the very end of the same year Fermi used the neutrino hypothesis to produce the theory of beta-decay and specify the form of the beta-spectrum as its consequence. When compared with the experimental data it showed that the mass of the neutrino was either close to zero or equal to it.

Its acceptance was not unanimous, however: the weightiest argument for the neutrino and, consequently, against non-conservation, belonged to the lower methodological level than the hypothesis of non-conservation. It did not use radically new ideas and seemingly was not a "new-generation theory" worthy of its predecessors – quantum mechanics and quantum electrodynamics. It did nothing more than justify the law of conservation and gloss over the deep-rooted problems with a physical constant that described beta-interaction and would be reduced to fundamental physical constants in future [148].

### *4.2.3. Non-conservation of Energy, General Relativity, Cosmology and Astrophysics*

Those who were not satisfied with the new experimental data and Fermi's theory were impressed with Landau's fundamental argument about incompatibility of non-conservation and General Relativity. In his letter to Bohr dated December 31, 1932, Gamow wrote:

> In the beginning of December I was at the Institute in Kharkov to look at the fast protons which they got there. Ehrenfest, Landau and some other theoreticians were also there, so we organized a

> small conference. We discussed many problems and cleared up one matter which I believe will be especially interesting for you. It looks as non-conservation of energy is in contradiction with the equations of gravitation for the vacuum. If the gravitational equations are correct for region $B$, this implies that the total mass in region $A$ ( where we do not know the laws) must be constant [there is a drawing in which region $A$ is shown like a small part of region $B$]. If in $A$ we have, for example, a $RaE$ nucleus, and alter its total mass with a jump in a transmutation process, we can no longer apply the usual gravitational equations in region $B$. In what way we have to change the equations is not clear, but it must be done. What do you think about this? [247, p. 568]

One can easily imagine how puzzled Gamow was – Bohr's hypothesis of non-conservation first appeared in his work published in 1930 [143]. (Bohr himself made it public in 1932.)

Ehrenfest came to Kharkov on December 14, 1932 and stayed there for a month [285, p. 152]. Bronstein was also there; Landau voiced his argument during a discussion of Bronstein's article [16]. The article preceded Bronstein by a month – it arrived at the editorial board of a journal that appeared in Kharkov in foreign languages. In his article "On the Expanding Universe", he brought together two fundamental themes: the time asymmetry of cosmology and relativistic quantum theory. They met at a point where Bronstein tried to create a cosmological model that would realize Bohr's hypothesis of non-conservation of energy. Being fully aware of the situation in relativistic cosmology and the potential of General Relativity that did not embrace quantum theory he believed that the cosmological problem in general, and the problem of time asymmetry in particular, could not be solved within GR. This problem could be solved only by a quantum relativistic theory. This meant that according to Bohr's hypothesis non-conservation of energy should be taken into account – Bronstein supposed that effectively this resulted in the cosmological term $\lambda$ of the GR equations depended on time.

This was the first physical constant of which the time dependence was connected with the expanding universe.[3] Cosmology today, based on the unified theory of interactions, also employs a cosmological constant that depends on the age of the universe (on its temperature, which changes with age). Just as in Bronstein's model the contemporary constructs allow energy to be moved from the "visible" forms of matter to the "invisible" $\lambda$ field. It seems that once made public the ideas took on a life of their own.

In a comment dated January 13, 1933 (prompted by the discussions in Kharkov with Landau and Ehrenfest, whom Bronstein thanked), which accompanied Bronstein's article, he first published Landau's argument:

> It was Landau who first attracted my attention to the fact that the gravitational equations of Einstein's theory for the empty space that surrounds a material body are incompatible with the non-conservation of its mass. This can be strictly verified by the Schwarzschild solution (spherical symmetry); in physical terms this is connected with the fact that the Einstein's gravitational equations allow only transverse, and not longitudinal, gravitational waves.

> The use of macroscopic instead of microscopic equations allowed to bypass this difficulty; the birth of radiation energy in the stars' nuclei [it was believed at that time that they were explained by quantum relativistic theory] is interpreted as a new energy form connected with $\lambda$-field. This compensates Bohr's non-conservation. This way out seems to be extremely unpleasant, though today there are no other ways. This is a puzzling paradox that is typical of the difficulties connected with the cosmological problem. [For more details, see Chapter 5.]

It is known that in General Relativity the mass of a spherically symmetrical source in an empty space cannot depend on time and that the transverse nature of the electromagnetic waves is connected with the law of charge conservation. As we can see, Bronstein used more definite terms to describe the incompatibility of the non-conservation hypothesis and GR than Gamow in his letter (and articles [147, 148]).

The following can be said about Landau's comment. In General Relativity, energy (or the mass that corresponds to it, $E = Mc^2$) is a source of the gravitational field. Therefore, just like in electrodynamics, it abides by Gauss's theorem (the total flux of a vector field through a closed surface is equal to the volume integral of the divergence of the vector taken over the enclosed volume). It was impossible to surmise a violation of the laws of conservation only in the microworld and pin all hopes onto a future quantum relativistic theory. After all, if a microvolume is placed inside a larger volume that is covered by GR the laws of conservation would have been violated within GR.

Even if Landau's argument was not quite definite from the mathematical point of view, for a physicist it nearly killed the non-conservation hypothesis. This was why Bohr responded with a desperate suggestion that the theory of gravitation did not apply to nuclear particles [118, p. 172]. This failed to save anything – the theory of gravitation should be changed even outside the microworld. Gamow had the following to say on this score: "a rejection of the law of energy conservation should inevitably change the general gravitation equations for empty space. This could be done but might cause a lot of trouble" [148, p. 391].

Being better versed in GR, Bronstein could better realize the full extent of this "trouble". In the introduction to his main work on quantum gravity he wrote that the circumstance Landau had pointed to

> probably excludes any possibility of a violation of the law of energy conservation in material systems even if they are outside the scope of GR (for example, in the systems within the "relativistic theory of quanta"). Indeed, any changes of energy, and hence mass, should cause gravitational waves in the surrounding empty space that lives within the common ("non-quantum") GR. The symmetry suggests that these should be longitudinal waves that is excluded by the gravitational equations for an empty space. Landau's qualitative argument – so far has failed to be formulated quantitatively [31, p. 196].

By late 1935 (when this was written) the mathematical indefiniteness of this interconnection had ceased to be important since the non-conservation hypothesis was no longer particularly attractive. Despite its indefiniteness and the ridiculously insignificant gravitational effects in microphysics the supporters of non-conservation took this argument harder than some of the new experimental findings on the beta-spectra. One has only to leaf through what Bohr and Gamow wrote at that time [118, 147, 148]. Later, in 1937, when the conservation laws were no longer a burning issue in the lecture he delivered when in the Soviet Union, Pauli described this argument as an important breakthrough [251].[4] Theoreticians were quite prepared to alter physical concepts to move their science forward, yet it was quite a different thing to discard the classical heritage of which GR had become a part. This is the usual combination of radicalism and conservatism in science.

The gap between non-conservation and gravitation in nuclear physics was not as wide as one might think. From the very beginning (when this hypothesis was first worded by Bohr in his manuscript of 1929 [247]), it was used to explain the source of solar energy; with time its astrophysical application became even more substantial. The main component in this was Landau's work of 1932 on the critical mass of a Fermi gas star [214]. Today this result is solely associated with the theory of white dwarfs and black holes; back in 1932 Landau believed that he discovered the domains (which he called pathological) that could be described by a $c\hbar$ theory and that were producing radiation energy out of "nothing" in accordance with Bohr's idea. This was a hint to a certain cycle in which energy was born: a pathological domain – a gigantic nucleus – which radiated high-energy beta-electrons and absorbed low-energy beta-electrons [81, p. 230].

Today one might be puzzled by the insistence with which the problem of the energy source in stars was tied with non-conservation. After all, this role could be played by the helium synthesis out of hydrogen and annihilation of the electron and the proton, not yet excluded by the laws of conservation of the lepton and baryon charges. Both ways of energy radiation by stars were well known and discussed; Bronstein also took part in such discussions.

Why then were they not favorably accepted? In the first place they both rested on the law of energy conservation ($\Delta E = \Delta Mc^2$) while in the special conditions of stars these laws might not be applicable; besides, the theoreticians were handicapped by their maximalism, with which they were handling the astronomic material. They wanted to explain everything by the basic principles only. The theory of stellar evolution led to a dead end: its key supposition about the ideal gas of which the stars were made was obviously oversimplified. (It was much later that a vast area of its applicability was discovered.) At the same time, observational material (the Herzsprung–Russell diagram) pointed to a one-dimensional evolutionary correlation between stars of different types and obviously demonstrated that a fundamental physical explanation was lacking. At that time the mechanism of burning (or synthesis), widely accepted today, was rejected because it produced too much helium "ash" and failed to account for the evolutionary transitions of the states of stars that differed greatly in mass.

Anybody familiar enough with the complicated (and far from one-dimensional) material concerning stellar evolution that has been accumulated since that time realizes that these were somewhat narrow-minded (or even naive) considerations. It is known that a star can explode in addition to radiating its mass away according to the relativistic law $\Delta E = \Delta Mc^2$. In the thirties, however, the arguments of the "pathological domains" were readily accepted; Ambartsumyan was one of those who treated them seriously [91]. By the mid-thirties he had drifted into astronomy very far from his physicist friends of the Jazz-band.

Astrophysics was not the only field to be affected by the conservation law problem; the cosmological motive in the story we are telling and the fate of Landau's argument testify that "experience as a supreme judge" alone cannot describe the evolution of a physicist's ideas.

In 1957 Pauli admitted that his antipathy to the non-conservation hypothesis had been rooted not only in an empirical fact (the upper limit of the beta-spectrum) but also in two theoretical considerations [252, p. 393]. First, while not doubting the law of conservation of an electrical charge, he saw no reasons why this law and the law of energy conservation should have different fundamentality levels (this was what Landau pointed out by drawing a parallel between the electrical and gravitational charges). Second, he did not admit that non-conservation of energy in beta-processes should imply the nonreversible nature of physical phenomena at the fundamental level. At the same time, this circumstance, viz. a possible time asymmetry of the $c\hbar$ theory attracted Landau and Bronstein to the non-conservation hypothesis – at that time they were much taken by the problem of cosmological irreversibility [22]. It is interesting to note that twenty-five years later, when a real violation of the law of conservation (of parity) caused a storm in physics, Pauli sought a connection between it and cosmology [252, p. 383].

Today, it seems to be inevitable that any discussion of the laws of conservation should invoke a connection between them and the space-time symmetries and, in particular, between the law of conservation of energy and the homogeneity of time. It is quite easy to supply arguments in favor of non-conservation if one bears in mind that the expanding universe, that is, non-homogeneous time in cosmology, was just recently established.

Today these connections, described by the Nöther theorem, are quite well known [126]. The records of the discussions held in the thirties contained only one such reference, however: in 1936, at the very peak of the Shankland crisis, B. Finkelstein reminded of such a connection. True, he referred to Jacobi's classical works[5] and didn't mention any association between the non-conservation idea and the cosmological time asymmetry. He pointed out that a possible violation of the laws of conservation would have brought about a radical transformation of the concepts of space and time in a fundamental theory of the future [264, p. 342]. Finkelstein was on staff at the Leningrad Physicotechnical Institute, he cooperated with Bronstein;

he was a specialist in the physics of solids, a field far removed from fundamental physics. Probably, the Nöther argument figured during oral discussions.

It is frustrating to realize how much was decided and discussed only orally and never reached publication or even private letters. Bronstein was an active correspondent though very few of his letters survived.

When writing about oral discussions and scientific publications one is tempted by the shop-soiled metaphor of an iceberg – a physical theory is propelled forward by some invisible underwater currents and the changing winds. The uninitiated would believe that the theory is advancing along predetermined paths that could not be changed although they might take unexpected and far-reaching turns. The underwater part may contain things that seemingly have little in common with science.

## 4.3. Non-Physical Arguments Applied to Physics

At all times science is affected by social-psychological factors determined by science's social status and by scientific-psychological factors determined by the variety of world-views present in physics. They affected Bronstein's stand on non-conservation.

In the Soviet Union, still rumbling from the recent radical social changes and boiling with ideological discussions, the non-conservation hypothesis was physics' hot spot. Relative domination by young scientists and a general conviction that the natural sciences were the key to the current technical revolution and, hence, to future socialist reconstruction added a cutting edge to scientific debates.

Landau, Gamow and Bronstein were among the front ranks of supporters of non-conservation. Landau made the largest contribution to its development: two of his results supported it and one did not. In 1931, Landau, together with Peierls, contributed work on the relativistic generalization of the indeterminacy principle, generally taken as a forecast of some future radical changes in $c\hbar$ theory. This was in accord with non-conservation, which common opinion related to the $c\hbar$ phenomena. Landau's work of 1932 on the critical mass of a star was taken as an outline of the domain where the $c\hbar$ theory resulted in non-conservation. These two works were outweighed by his counter-argument – the incompatibility of GR and non-conservation. Gamow was the first to make Bohr's hypothesis public, even before its author deemed it necessary to do this. He repeatedly supported it in his surveys and popular articles. Bronstein seemed even more keen on the hypothesis – he discussed it in two popular books [81, 82] and an article [79].

In the thirties the discussion spilled onto the newspaper pages, where over-zealous journalists anathematized everyone who went against the conservation laws. A certain V. Lvov who called himself a scientific journalist was especially active.[6] He was smart enough to deduce the equation $E = Mc^2$ from philosophical quotations. In contrast to many of his colleagues, he was enthusiastic about the theory of relativity but rushed in to blame its idealistic distortions among which he classed cosmology; much later, in 1958, he wrote the first Soviet biography of Einstein, which appeared in the "Illustrious Lives" series. Back in the thirties he was calling to "do away" with a group of physicists headed by Landau and Bronstein who knew no restraints in their defense of the non-conservation hypothesis and were forcing Soviet science back to the stone age [227–229]. (Three decades later, under different political circumstances, he found different words for the same people [231, 172].

Indeed, Lvov went a little too far in defending the laws of conservation[7]; this was the time when even articles written by physicists were affected by ideological heat and non-physical arguments. Such was an article by D. Blokhintsev and F. Galperin of 1934 [110]. The arguments in favor of the "great, eternal and absolute natural law" were mostly ideological; the authors did their best to distort the position of the "zealous persecutors of the law of conservation of energy in the country of dialectical materialism (Gamow, Landau, Bronstein and others)" [110, p. 106]. They branded the position of the non-conservation supporters as idealist and attributed it to the flaws of "propaganda of dialectical materialism among physicists". There was a long list of cases where this law was evident in atomic physics – actually, the only physical argument in its favor. One would think that Bohr was familiar with it – suffice to say, his formula $E_2 - E_1 = h\nu$. The authors [110] suggested several ways out of the crisis brought about by the continuous beta-spectrum that would preserve the conservation laws (the neutrino hypothesis came last with them).

This rage of non-physical arguments urged Ioffe who, being an experimentalist, did not sympathize with Bohr's hypothesis and was keeping away from the discussion, to speak in defense of the purely physical side of the controversy

> For some reason the very discussion is seen as a crime against dialectical materialism. For my part I am sure that this is caused by a complete ignorance .... There is not a single experimental law that could claim validity for a domain first covered by experiment. There are no sacred laws in physics, and there can be no such laws. The law of conservation is not sacred, and there are absolutely no grounds for canonizing it [195, p. 60].[8]

Was it in the heat of discussion that Bronstein used non-physical arguments to defend Bohr's hypothesis? A careful analysis of what he wrote at the time and interviews with those who knew him suggests that he simply answered in kind to bring out the absurdity of such arguments [79].

One can hardly say that Bronstein could see no further than physics – he was interested in philosophy and looked deep into the dialectics of developing knowledge. Naturally enough, he was disturbed by the clumsy attempts to prove a physical point with quotations from philosophical treatises. In his book [82] he looked to history to support his doubts about the universal applicability of the laws of conservation; he then switched to the situation brought about by the Ellis–Wooster experiments. He pointed out that the transformation of elements, which had been taken to be impossible after many centuries of alchemists' futile attempts, turned out to be feasible after all in nuclear physics. His main idea was that while being definite enough to channel human thought philosophy was not definite enough to turn a physical result into an absolute.

There were enough people who thought, together with Lvov, that "the Marxist-Leninist teaching acts as a supreme arbiter in the Ellis–Wooster experiment just like anywhere else" [225]. Lvov was bold enough to insist that the supreme arbiter had already passed judgement on everything and that these judgements could be found in the thick volumes. In the late forties and early fifties these arguments ruined biology in the Soviet Union. In the physics of thirties such attempts were actively rebuffed. Bronstein was quite right to defend the non-conservation hypothesis from the philosophical point of view – after all, its death belonged to physics, not philosophy.

More often than not non-physical arguments are invited by physicists overeager to prove their point – this is not the exception in scientific discussions as one might imagine. In the physics intuition sometimes brings the mind outside logical deductions and inductions. Science is not a multiplication table – there are frequent situations when intuitions and research programs, far outstripping experimental limits, cross swords. With no physical arguments left, the physicist turns to his entire cultural baggage where the choice depends entirely on his world perception.

Here are two situations when non-physical arguments were invoked. In the famous discussion on quantum mechanics they spoke about consciousness and justice, said that God did not play dice etc. [169]. Sergei Vavilov supplied his 1928 book on relativity (the experimental basis of which was not particularly stable at the time) with a quotation from Newton. His obvious purpose was to prop up the theory of relativity; taken out of context Newton's words acquired a different meaning [129].

In any case, a non-physical argument betrays a strong bias on the part of the author. There is no doubt that Bronstein was partial to non-conservation. To understand his motives better, let us look at the problem's scientific and psychological dimension, the research programs and the differences in the physicists' world-views.

One can differentiate between two types of theoreticians that could be called "thinkers" and "pragmatics", or "doers". They differ radically in the character of the problems they take up for consideration and the intuitive assessment of a situation. The pragmatics believe, to borrow an expression from Landau, that human life is short enough to stop them seeking answers to problems that cannot be promptly sorted out. The thinkers, on the other hand, do not regard physics as a body of assorted problems – they feel an integral picture of the world should be their prime necessity.

Taken in their pure forms both types can be depressingly boring – in actuality most theoreticians combine different features. One can speak only of a predominance of one type over the other: Einstein, a thinker, was not a stranger to innovation and invention [288]. Scientific achievements and their scales do not directly depend on the type of world perception – one may even formalize the situation in the following way: if we take $T$ for thinker component and $P$ for pragmatic one then creative ability would be determined by $(T-P)$ while scientific achievements by $(T \times P)$.

Outstanding physicists belonged to both groups. Their cooperation was indispensable in moving science further on. In a certain sense thinkers find life harder than their pragmatic colleagues because they are dealing with a much larger edifice. The history of physics testifies, however, that it was precisely the thinkers who produced the greatest changes in the physical picture of the world.

Like Bohr, Bronstein was undoubtedly a thinker; this explains why it took them longer to abandon non-conservation; in fact, Bohr abandoned it in as late as 1936. The reasons that forced them to accept the hypothesis were much more fundamental than Fermi's theory spearheaded at an isolated physical phenomenon, beta-decay.

If he were a pragmatist, Bohr would never have pushed forward the hypothesis that had been experimentally rejected some years earlier. It is quite natural, though, for a man searching for a picture of the world into which the hypothesis fit.

As for Bronstein, his adherence to non-conservation can be explained by the nature of his interests. He never limited himself to microphysics and displayed a lively interest in the fundamental problems of astrophysics and cosmology. It was his conviction that they needed the $c\hbar$ theory (for more details, see Chapter 5). At that time the non-conservation hypothesis was always associated with a coming consistent $c\hbar$ theory.

## 4.4. A Duel in *Sorena*

To obtain a clearer picture of how views differed on Bohr's hypothesis, let us have another look at Bronstein's article "Is Energy Conserved?", which appeared in the first issue of *Sorena* (a Russian abbreviation for *Socialist reconstruction and science*). It was the most scientific of all popular journals of the time. Its editorial council included many prominent academics and Nikolai Bukharin was its editor-in-chief. The editorial board deemed it necessary to counterpoise Bronstein's article with a critical contribution from S. Shubin, "On the Conservation of Energy", and to invite others to contribute to the discussion. Indeed, the problem was the hot issue in the mid-thirties.

In his summary of the fifty years of Soviet theoretical physics, Tamm referred to Bronstein and Shubin as "the remarkably bright and promising physicists of their generation" [268]. Both were arrested in 1937 and perished before they could realize their creative potential.

Semen Shubin (1908–1938) studied under Mandelstam and Tamm. In 1932 he was appointed head of the theoretical department of the Urals Physicotechnical Institute and contributed a great deal to set up a physical school in the Urals [136, 137]. S. Vonsovsky, who took over the school of magnetism after him, dedicated his definitive book on magnetism to the memory of his teacher and friend [135].

To pin down the differences in treatment of the conservation laws, one has to take into account that Shubin was mainly occupied with the physics of solids. He had to suspend his university studies for a year because of his active political stand (in 1928 he was exiled to the Urals); in 1930 he volunteered at the construction site of Magnitogorsk where he was working on a newspaper. He was genuinely interested in ideological problems, including the relationship between Marxist philosophy and the physics of his time. One can clearly see this from Shubin's manuscript materials on the methodological and philosophical analysis of quantum physics.[9]

He seemed to regard the "mechanists of Timiryazev's ilk", who banned new physics in the name of dialectical materialism, as main foes. He first encountered this philosophy as a student at Moscow University where Timiryazev was a professor. In his manuscript Shubin didn't doubt that "the future specified formulation of the laws of the microworld" that ruled inside the nucleus and were related to quantum theory would bring an even more radical change than quantum mechanics; he supported his point by Landau and Peierls' work of 1931.

Here Bronstein and Shubin agreed; it should be said that nearly all physicists shared this view – they would have been bitterly disappointed had they realized that the desired changes would take much more time.

Why, then, did their views of the conservation laws differ so?

Let us consider the details of the *Sorena* duel. While giving Bronstein his due where the physical (experimental and theoretical) arguments were concerned, Shubin very sarcastically interpreted about his philosophical considerations:

> There is no direct experimental evidence for or against the law of conservation of energy in nuclear physics. Neither are we aware of any theoretical considerations that would allow us an unambiguous solution of this problem since there is no relativistic theory of quanta. Being materialist dialecticians we are wielding a mighty methodological weapon that allows us to look the future in the eye. This principle says: "everything can happen". The law of energy conservation so dear to the heart of a bourgeois bookkeeper building his world in accordance with his carefully kept accounts can be toppled any day now. The eternal dream of a *perpetuum mobile* may come true in communist society.

Shubin, of course, was pushing things too far. In actual fact, Bronstein's exposition was vivid, if a trifle superficial. He pointed out that the dominant philosophy was very much dependent on the prevailing social reality and outlined how the ideas of a *perpetuum mobile* and of energy conservation were changing with time. The only conclusion that he drew from materialism was that "no physical law can be regarded as a dogma or an a priori absolute and universal truth".

Speaking on behalf of Marxist physicists, Shubin appreciated "the fact that such a person as Bronstein took up materialism to support his arguments" and added, not without malice, that he seemed to suppress "out of modesty what others said, much more definitely and firmly, about the law of conservation of energy in Marxist publications. One of these others was a rather well-known Marxist, Friedrich Engels".

Indeed, Engels' pronouncements were quite well known at the time and frequently employed by the opponents of non-conservation to boost their arguments. Lvov was one of those who invoked Engels to rule out any doubts about the law of conservation (though ignoring the difference in the physics of the two ages). Naturally, Shubin never ruled out anything. He offered the following formulation of Engels' position: the very existence of the laws of conservation just "reflected a very general fact that motion could not be destroyed and that, therefore, the laws should be reflected, *in one form or another*, in *any* correct physical theory".

Naturally enough, Bronstein never tried to deduce the non-conservation hypothesis out of materialism:

> As we have seen, materialist dialectics teaches us that the laws of conservation *might* prove to be wrong, but it has never asserted that it *should* be wrong in the quantum relativistic sphere.

He employed this philosophy to promote his point rather than to prove it. The task was to show that the laws of conservation were not infallible – being a profound physicist and having a deep knowledge of the history of physics he was aware of the psychological barrier that blocked non-conservation. After all, in 1930 when writing about the doubts in the conservation principle caused by the discovered radioactivity, it was Bronstein who called it "one of the key and most reliable of all physical laws" [63, p. 25].

Now let's turn to their physical arguments. On the one hand, Shubin pointed to Fermi's theory as a hint of the conservation laws' potential, while Bronstein didn't mention it; on the other hand, Shubin ignored Bronstein's considerations on the limited nature of concepts in the relativistic quantum sphere and on "the impossibility of cutting up an electron".

Why did Bronstein leave Fermi's theory out? Did he fail to appreciate it? His review of the advances of nuclear physics that appeared in the very next issue of *Sorena* described Fermi's work as the greatest achievement of the year [80]. At the same time Bronstein believed that this was a tactical rather than a strategical advance – and he was in good company. This was what Fermi himself was thinking at the time; Pauli was even more skeptical. For them, the theory's outward perfection did not compensate for its internal shortcomings. It would be hard to explain these troubles in succinct popular terms as *Sorena*'s style demanded. How could anybody not on the best of terms with nuclear physics believe that vague considerations about the quantum relativistic uncertainty were preferred to verifiable calculations? This was why Bronstein left Fermi's theory out.

The different stands taken by two physicists on non-conservation were predetermined by their research programs – Shubin was mainly engaged in the applicability of quantum mechanics to solid state physics; it was not for him to look for possible faults in the building's foundation. By contrast, Bronstein was working in fundamental theoretical physics; it should be said that the quantum relativistic considerations that Shubin had preferred to ignore acquired a more definite and fundamental shape in Bronstein's work on quantizing gravitation that appeared in the same year (1935; for more details see Chapter 5).

As often happens in a discussion between two talented and earnest people, both were correct to a certain extent. Shubin was right when he insisted that a non-conservation hypothesis could not be deduced from philosophical deliberations and that by 1935 nuclear physics had left little hopes for "non-conservators". In his turn, Bronstein was right when he said that physics would be unable to bypass the quantum relativistic problem and that the laws of conservation could not be proved by philosophical arguments. Were Shubin better acquainted with ironic Bronstein, he would have suspected that an argument of the "bourgeois nature" of the conservation laws was prompted to shame opposite – and much more often – non-physical arguments.

## 4.5. The Death of a Non-conservation Hypothesis

Today it is very hard to recreate the atmosphere in which the conservation laws could be doubted. One can even think that by 1935 it was the stubborn and not very bright physicists who were still holding onto the non-conservation hypothesis. In actual fact, this was the time of stormy developments in physics akin to the Newtonian age – the conservation of momentum (true both in physics and psychology) held the promise of more miracles. To throw the futility of the non-conservation hypothesis into bolder relief, Shubin likened it to the *perpetuum mobile*. This argument cut both ways: Bronstein reminded his readers that another futile alchemist dream of the transformation of elements had recently come true [82].

The way the physical community responded to Shankland's experiments was even more illustrative. By early 1936 the idea of the neutrino (and the conservation laws) had been empirically pinned up and advanced in theoretical terms, while the non-conservation arguments were shrinking before the physicists' eyes until completely incapacitated by their incompatibility with GR. The following response to Shankland's assertion that photon theory and the conservation laws did not apply to the Compton effect in the gamma-range seem all the more puzzling.

> Thus physics is now faced with the prospect of having to make a drastic change in its fundamentals, a change involving the giving up of some of its principles, which have been most strongly relied on (conservation of energy and momentum), and the establishment in their place of the Bohr, Kramers and Slater theory or something similar. The only important part that we give up is quantum electrodynamics. Since, however, the only purpose of quantum electrodynamics, apart from providing a unification of the assumptions of radiation theory, is to account for just coincidences as are now disproved by Shankland's experiments, we may give it up without regrets – in fact, on account of its extreme complexity, most physicists will be very glad to see the end of it.

When was this written? Who wrote it? One may find it hard to believe that it was in February 1936 that Dirac, one of the founding fathers of quantum electrodynamics, wrote it [180]. According to a contemporary survey "The weightiest argument in Shankland's favor [was] the fact that the experiments were performed under Compton, who thus shares responsibility for the results" [300].

"Physical laws should have mathematical beauty", true to this credo Dirac was concerned with the lack of a consistent (and mathematically beautiful) relativistic quantum theory.

Experimental results obtained in a well-known laboratory and supported by one of the prominent theoreticians sapped the faith in the conservation laws. The materials of the 1936 session of the Academy of Sciences provide ample evidence of the sentiments of the Soviet physical community [264]. Ioffe stood practically alone against Shankland's experiments. Others were more or less inclined to accept them.

The above mentioned survey ended with the words "even if the laws of conservation are violated, the philosophical prerequisite of the non-destruction of motion naturally holds. If the invariants applicable to macroscopic motions and totally applicable to the microscopic processes with heavy particles are found not to fit with the photon-electron interaction, this should be taken as a stimulus to look further for new, more general, invariants".

It should be pointed out that the enthusiasts of non-conservation gave a cool reception to Shankland's experiment [84], since it offered no *theoretical* support for the hypothesis; therefore, as soon as the results were experimentally disproved the non-conservation hypothesis died a natural death.

Is is worth writing about a hypothesis that was disproved and forgotten? A true appreciation of scientific advances requires a broad background with all the pain and anguish it sometimes causes. Besides, a true idea is not always fruitful, just as an incorrect idea sometimes bears much fruit. A closer look at the changing fortunes of the conservation laws in the twenties and thirties revealed an intricate pattern of facts and ideas present in a theoretician's mind and the role of his world perception and research program.

We have isolated the events connected with the non-conservation hypothesis to let them stand in bolder relief. In actual fact, throughout these years, Bronstein was also studying semiconductors, astrophysics, nuclear physics, the equilibrium of radiation and pairs under super-high temperatures and the relativistic generalization of the uncertainty principle. In the summer of 1935 he became interested in quantizing gravitation; several months of strenuous effort resulted in a doctorate and two articles. It seems that the quantum gravitation, or $cG\hbar$-physics, affected his stand on non-conservation. He traced the correspondence between $cG\hbar$- and $cG$-descriptions, therefore, the incompatibility of non-conservation and General Relativity became glaring. His note in the article dated December 1935 [31] (before Shankland's results became known) showed that he had cooled to the idea of non-conservation.

# Chapter 5
# $cG\hbar$ Physics in Bronstein's Life

Bronstein's main contribution to physics were his works on the theory of gravitation and cosmology. His thesis "Quantizing Gravitational Waves" written in 1935 proved to be closer to modern fundamental physics than to the physics of his own day.

## 5.1. An Unsuitable Thesis

Back in 1935 there were enough reasons to raise brows at the quantum theory of gravitation as an obviously unsuited research project or the subject of a doctorate thesis. First, it was not a topical issue: Einstein and his few supporters were busy deducing quantum laws from the unified field theory that was expected to generalize General Relativity. The rest of the physical community was engrossed in the theory of the nucleus and quantum electrodynamics. The domain of phenomena, later called elementary particle physics, was waiting for their theoretical substantiation; gravitation was left out in the cold. In the thirties Bronstein's subject remained outside the mainstream of physics.

True, General Relativity itself was hardly topical when it was first formulated. There were deep-seated reasons behind it born out by the situation in physics: the earlier theory had proven to be inadequate [127]. Where its external quantitative effects were concerned, the relativistic theory of gravitation was no match for Newtonian universal gravity and other fundamental theories as they stood when first knocked together. Actually, physics felt no pragmatic need for a quantum theory of gravitation. The history of physics, however, abounds in significant practical results stemming from the theories not needed by anybody except few theoreticians.

Here again we can take up the fundamental constants to describe physical theories. Then Newton's gravitation theory can be called a $G$-theory, General Relativity a $cG$-theory, while the quantum theory of gravitation can be called a $cG\hbar$-theory. This terminology, which can be traced back to Bronstein, help to reveal the structure of theoretical physics.

In the thirties, the $cG\hbar$ theory as a thesis subject promised nothing new either practically or theoretically. In any case, in 1929, Pauli and Heisenberg, the two greatest authorities in quantum physics, declared: "We should also say that the quantizing gravitational field, which is indispensable for certain physical reasons, can be done quite easily with the help of formulae totally analogous to that which has been developed here" [159, p. 32]. An important comment is needed: they implied that it was a weak gravitational field that was being quantized; it could be amply described by Einstein's linearized equations. This was precisely how L. Rosenfeld treated the subject in 1930 [259]. Although evoking an analogy with electromagnetism this obvious approximation left out specific properties of gravitation determined by the equivalence principle and the field's geometrical and non-linear character. It was Bronstein who first discovered the fundamental complexity of quantum gravity.

His colleagues were taken aback – after all, his results in the physics of semiconductors were well-known and the subject's promising nature was obvious to everyone. Ioffe, director of the Physicotechnical Institute, himself described them in flattering terms [193]. Y. Frenkel went so far as to write: "Today his doctorate on semiconductors (electronic semiconductors) that he will defend in the near future is practically ready" [173, p. 322]. On June 10, 1935, the Institute's Learned Council, having granted Bronstein his candidate degree for his work on astrophysics suggested that he could present a doctorate on the theory of semiconductors. It required uncommon boldness to choose another, totally different, subject for his doctorate, but he preferred this rather than go back to a well-chewed theme.

Some short historical commentary is needed here. Before 1934 there were no academic titles in the Soviet Union: they had been abolished together with military ranks after the 1917 revolution. In science a reliable system of command and discipline is not as prominent as in the army. It favors rather a flexible structure and unhampered initiative. Indeed, can one fit scientific achievements into a rigid pattern? Is there any need to do this? Back in the thirties young Russian physicists were skeptical about this. However, the Soviet government, "in order to boost up research activity and upgrade the level of research associates and lecturers", re-introduced academic ranks in a two-level form: a candidate and a doctor.

Naturally enough, a certain number of doctors was needed to set the machine going; members of the Academy of Sciences were granted doctorate degrees. Besides, some of the "lesser fry" (for instance, heads of theoretical departments of the newly-established Kharkov and Sverdlovsk physical institutes Landau and Shubin) got their doctor's degrees without defending theses. Bronstein, who stood lower on the administrative ladder and remained in Leningrad among many other bright physicists, had to go through the circuit.

As everywhere else throughout the world the young and talented physicists treated the academic ranks without much reverence. It took the director and sector heads of the Leningrad Physicotechnical Institute a lot of effort to stir up their younger colleagues to write and defend theses on time before the end of 1935. According to I. Kikoin, who witnessed a regular dressing down in the director's office, Bronstein put up a formidable defense by insisting that a doctorate without long formulae would not be imposing enough and that therefore semiconductors were a poor choice. It was a joke that had a grain of truth in it: he had already been engrossed in quantizing gravitation, which indeed required extensive formalization. In view of this Kikoin came to Bronstein's presentation of the thesis supplied with a telescope to be sure not to miss any of his indices on the blackboard.

Bronstein completed his work during the summer of 1935: on June 10 the Learned Council was still expecting a thesis on semiconductors, when in August Bronstein submitted an article on quantum gravity [30]. Bronstein defended his thesis on November 22, 1935. His opponents, Vladimir Fock and Igor Tamm, two prominent theoreticians, were lavish with their praise. The stenographic report and eyewitness accounts show that it was more of a lecture than a proper defense procedure – after all, Bronstein's prestige stood high in the academic community [173].

What led him to the theme of his thesis?

Even as a student, he was highly independent in selecting research subjects. Besides, his superior, head of the theoretical department, Frenkel was very skeptical of quantum gravity; his article "The Principle of Causality and the Field Theory of Matter" is an ample evidence of this (the article was written, but not submitted, for the 1949 Schilpp collection dedicated to Einstein). There was a section dealing with "nuclear and gravitational fields". The author disputed an opinion that "the gravitational field or, at least, the weaker part, which forms gravitational waves, could be quantized with the corresponding particles – gravitational quanta or gravitons – identified". He wrote further: "Einstein was probably the first to assimilate gravitational waves and the corresponding particles in a conversation with the author back in 1925. A detailed mathematical investigation of the problem was first published in the Soviet Union by Matvei Bronstein in 1936.".

For Frenkel an analogy between the gravitational and electromagnetic fields was too superficial. He argued that while "the electromagnetic field [was] *matter* while the gravitational field merely [determined] the metrical properties of the space-time continuum." He insisted that "strictly speaking there [were] no such things as gravitational energy or gravitational momentum since the corresponding values [did] not form a true tensor and [were] nothing more than a *pseudo*tensor." It was his conviction that this was responsible for the failures to formulate unified field theory for both fields. Frenkel assumed that the attempts to quantize gravitation were senseless since "the gravitational field [had] a macroscopic, rather than microscopic, meaning; it merely [supplied] a *framework* for physical events occurring in space and time while quantizing [could] be applied only to microscopic processes in *material fields*". These considerations were little affected by the developments in physics that occurred after November 1935 – this and his remark during Bronstein's defense [173, p. 319] allows us to surmise that his position was the same in 1935.

This controversy gives us an insight into the climate of tolerance that was reigning at that time in the theoretical department; on the other hand, it shows that quantizing gravitation was far from merely a technical task and an obvious and straightforward problem.

Yakov Frenkel's interest in Einstein's theory of gravitation was deep-rooted: he was the first in the Soviet Union to offer a systematic exposition of General Relativity in which he betrayed his bias to a specific electromagnetic picture of the world [130]. His stand on gravitation demonstrated his doubts about a synthesis of quantum and general relativistic ideas caused by a specific (geometrical) nature of the gravitational field, its identification with the space-time metrics, and also by obviously negligible gravitational effects in the microworld.[1] It was in the sixties that Rosenfeld voiced his doubts on the necessity of quantizing gravitation since, he argued, it was probably of a purely classical microscopic nature [260].[2] And this despite the fact that Rosenfeld was the first to formalize quantizing gravitation [259].

Bronstein never doubted the fundamental nature of gravitation. He believed that sooner or later physics would have arrived at quantum gravity and a complete $cG\hbar$ theory. At the same time, it was he who discovered why an analogy between electromagnetism and gravitation was not valid. This explained why the attempts to create quantum gravidynamics similar to quantum electrodynamics, merely by substituting the graviton for the photon could never be completely successful. He believed that it was necessary to quantize gravitation but that a complete theory patterned according to electrodynamics was impossible. He also obtained answers to the key questions in quantum gravity for the cases when the $cG\hbar$ effects were weak and the "electrodynamic" quantization pattern could be applied. These questions related to the correspondence between $cG\hbar$-theory and the $cG$- and $G$-theories of gravity, that is, the correlation between the quantum theory of gravitation, General Relativity and the Newtonian law of universal gravity.

## 5.2. The Roots of Bronstein's Interest in $cG\hbar$-Physics

### *5.2.1.Quantum Gravity before Bronstein*

Einstein was the first to declare that gravitation needed a quantum theory. It was in 1916, several months after he finally formulated General Relativity, that, while discussing gravitational waves, he stated

> Due to electron motion inside the atom, the latter should radiate gravitational, as well as electromagnetic energy, if only a negligible amount. Since nothing like this should happen in nature the quantum theory should, it seems, modify not only Maxwell's electrodynamics but also the new theory of gravitation [305, p. 522].

Later, in 1918, he commented on his formula on the radiation of gravitational waves

> As one can see from the formula, the intensity of radiation cannot be negative in either direction; even less so can the complete radiation intensity be negative. I have emphasized in my recent paper

that the final result, according to which the bodies lose their energy due to thermal excitation, breeds doubts in the theory's universal applicability. I believe that an upgraded quantum theory would involve changes in the gravitational theory [307, p. 642].

Einstein's remarks relates to the famous problem of the atom's electromagnetic instability. However, his forecast, fortunately, lacked quantitative values. Radiating electromagnetic energy away leads to the electron falling down on the nucleus within the characteristic time

$$\tau_e = E/\dot{E}_e \approx mv^2/(\ddot{d}^{\,2}/c^3) \approx c^3 m\omega^{-2} \approx c^3 m^2 r^3 e^{-4} \approx 10^{-10}\, sec$$

which is a glaring contradiction to observations. Radiating away atomic energy in the form of gravitational waves (calculated by Einstein's formula) takes place with the characteristic time

$$\tau_g = E/\dot{E}_g \approx mv^2/\,(G\dddot{D}^{\,2}/c^5) \approx c^5 G^{-1} m r^4 e^{-4} \approx 10^{37}\, sec = 10^{30} \text{ years!}$$

Evidently, there is no contradiction with the empirical data. It seems that Einstein had in mind an analogy with electromagnetism.[3]

For two decades after Einstein had pointed to a necessity for a quantum theory of gravity there was no enthusiasm over it: the physicists were facing a challenge of other, more urgent, problems of quantum mechanics and quantum electrodynamics. The few and far between observations overemphasized an analogy between gravitation and electromagnetism.

In 1929 Heisenberg and Pauli wrote in [159] that quantizing gravitation would not produce fundamentally new problems compared to electrodynamics. They were convinced that quantum gravitation was necessary and referred to Einstein's remark; Klein in 1927 pointed out also, that there was a need for a unified description of gravitational and electromagnetic waves that would take Planck's constant into account [203].

The same observation seemed to prompt Heisenberg to ask whether infinites were inherent in quantum electrodynamics irrespective of the "electron problem", that is, whether they emerged in an absence of charges if a proper gravitational interaction of electromagnetic waves was taken into account. Rosenfeld tried to answer this question in 1930 [259].[4] He looked at a system that comprised an electromagnetic and a (weak) gravitational field, the interaction between which was determined by Einstein's linearized equations. This approximation (which Einstein obtained in 1916) allows one to ignore the geometrical nature of gravitation and the space-time curvature and to consider only two fields (vector and tensor) in flat space-time. By quantizing them according to Heisenberg-Pauli, Rosenfeld confirmed Heisenberg's supposition about the infiniteness of gravitational energy and supplied the first approximation for transmutations of the light and gravitational quanta. During Bronstein's defence Fock and Tamm emphasized, however, that Rosenfeld's results were of a purely formal nature and never suggested meaningful physical conclusions [173].

This was the gravitational context into which Bronstein plunged. Like his predecessors he looked mainly at the weak field but also offered an analysis of the fundamental difference between quantum electrodynamics and the quantum gravity not limited by the field's weakness or "non-geometrical" character. This analysis demonstrated that classical Riemannian geometry and the usual quantizing pattern were inadequate for a complete theory of quantum gravitation. It also revealed the limits of the quantum-gravitational domain.

### *5.2.2. "The Relationship Between the Physical Theories and Their Relation to the Cosmological Problem"*

Bronstein's interest was not prompted by an analogy between electromagnetism and gravitation and the possibility of expanding the methods of quantum electrodynamics onto a new sphere. He was mainly concerned with the general physical picture of the world and its integrity – a problem that had been interesting to him since an early age. Indeed, his very first works dealt with quantum and relativistic physics: article [1] discussed the quantum structure of electromagnetic field, while his work [4] displayed no mean knowledge of the General Relativity apparatus.

He deemed it necessary to round out his popular work *The Structure of Atom* (1930) with an overview of the contemporary state of fundamental physics. He mentioned Dirac's theory, quantum electrodynamics and Eddington's plans for a fundamental theory. He also wrote: "It is a task for the nearest future to identify the ties between quantum mechanics and the theory of gravitation".

Back in 1930 this prediction seemed rather far-fetched. There were many physicists who would object to this approach to physics's priorities even if they agreed that some connection did exist between gravitation and quanta – few were able to discern the phenomena that could be studied through this connection. On the other hand, for the group of physicists, led by Einstein, who were working on a unified field theory, the words "connection between gravitation and quanta" simply meant that "quanta would be deduced from the generalized gravitation theory". Bronstein did not share this opinion: he concluded the section on unified field theory in his encyclopedic entry [37] with the words: "it seems that Einstein's program of unified field theory will prove impossible" and that "what will be needed is a sort of a marriage between relativity and quantum theory". His profound knowledge of both showed him that they were equally fundamental and that their synthesis rather than subordination was desirable.

His popular article [60], which allowed more freedom of expression ended with the words

> Physics of the future will hardly tolerate the strange and unsatisfactory division into quantum theory (microphysics that supervises nuclear phenomena) and the relativistic theory of gravitation that governed macroscopic bodies rather than individual atoms. Physics will not be divided into microscopic and cosmic physics – it should, and will, be united and undivided.

Today, when particle physics is actively involved in cosmology, these words seem to be rather commonplace. What did they reveal for the thirties? They were not enough to assure his concern for quantizing gravity, but they betray Bronstein's interest in astronomy – these were the times when cosmology belonged to astronomy rather than to physics.

He had been deeply engrossed not only in the physical picture of the world but also in what can be called "a map of the world of physics". In 1933 he presented his views in the article "On the Possible Theory of the World as a Whole" in the section "Relationship Between Physical Theories and Their Relation to the Cosmological Theory". Later he reproduced the same material twice [50, 81] – a sure sign of the importance he attached to the subject.

What can be said about his map of physics? First of all, it was developing science – he used the future, past and present tense to describe it. Second, just like any geographical map this map had boundaries. Both "spatial" and "temporal" structures were determined by three constants, $c$, $G$, and $\hbar$. The boundaries divided the spheres of applicability of the fundamental theories that did not take into account some of the universal constants, while evolution meant new physical theories that would organically embrace these constants.

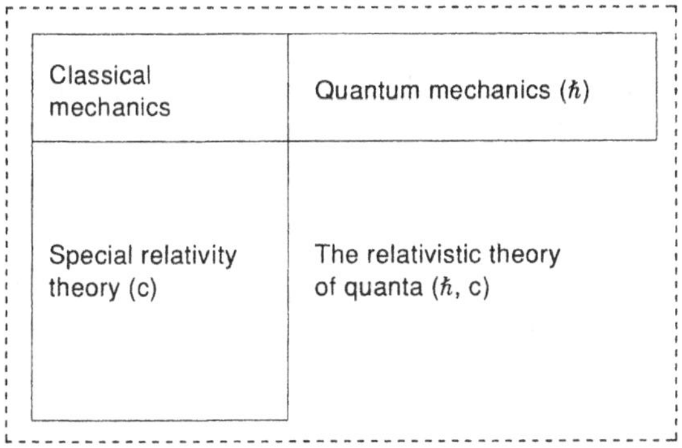

Classical mechanics
Quantum mechanics ($\hbar$)
Special relativity theory (c)
The relativistic theory of quanta ($\hbar$, c)

Fig. 1. "The areas of aplicability of quantum mechanics and special relativity theory intersect in the area of classical mechanics; the dotted line shows the area of applicability of the not yet created 'relativistic theory of quanta'" [21, p. 22].

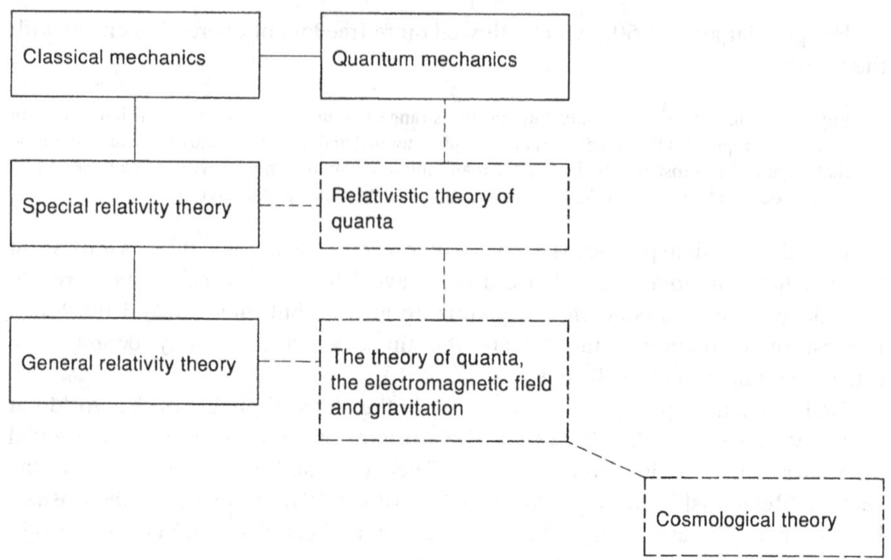


Fig. 2. Correlation of the physical theories and their relation to the cosmological theory; "continuous lines denote the already existing physical theories while the dotted lines correspond to the still unresolved problems" [21, p. 25].

Bronstein offered Fig. 1 to illustrate the correlation of classical mechanics, quantum mechanics, special relativity and the "relativistic theory of quanta" that needed to be evolved. Having introduced the constant $G$ by the means of General Relativity, he offered the extended Fig. 2 "embracing all physically meaningful questions that could be formulated today or, probably, all physically meaningful questions" [21, p. 25]. One can see that the nearest task was a relativistic quantum theory, the $c\hbar$ theory. He further explained why "the question of the values of world constants that have dimensions is physically meaningless" and wrote: "If a theory can explain dimensionless constants its task would be more or less complete, since the values of these constants are responsible for the specific picture of the world". He illustrated his thoughts with one of the problems of the $c\hbar$ theory – an explanation of the dimensionless constant of the fine structure $\hbar c/e^2$ . This explanation would have led to an explanation of $e$ (the electron charge) through constants $c$ and $\hbar$. True, at that time this was rather commonplace, yet the next forecast was rather unexpected:

> After relativistic quantum theory is formulated, the next task would be to realize the third step, shown in Fig. 2, namely, to merge quantum theory ($\hbar$ constant) with special relativity ($c$ constant) and the theory of gravitation ($G$ constant) into a single whole.

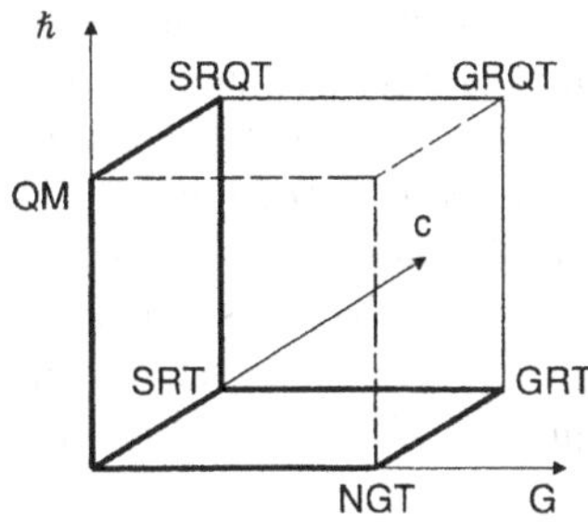


Fig. 3. The "space" of the physical theories in the cGħ system of coordinates.

NGT – Newton's gravitation theory
SRT – Special relativity theory
QM – Quantum mechanics
GRT – General relativity theory
SRQT – Special relativity quantum field theory
GRQT – General relativity quantum theory

(The non-trivial nature of the *cGħ* pattern is obvious when compared with Pauli's article of 1936 [250], where he employed the same constants to discuss the situation in physics.)

To illustrate Bronstein cited a problem for the *cGħ* theory to explain the dimensionless quantity $\hbar c/Gm_e^2 = 10^{45}$ or, by the same token, the electron mass $m_e$ through the $c$, $G$, and $\hbar$ constants. But he believed that cosmology was the main battleground for the *cGħ* theory:

> the solution of the cosmological problem demands as a preliminary step a unified theory of electromagnetism, gravitation and quanta which is shown in Fig. 2 with the second dotted-line rectangle [21, p. 28].

(To sound completely up to date this statement needs only fundamental interactions unknown back in 1933.)

This map of the *cGħ* physics can be found in Bronstein's article of 1933; he voiced the same ideas in February, 1932, when Frenkel's paper "On the Crisis in Contemporary Physics" was discussed at the institute [292].

The only change that time demanded of Bronstein's map was a switch from the two-dimensional to the three-dimensional presentation. A closer look reveals a certain omission in Fig. 2 – it lacks the Newtonian theory of gravity and the path from the *G*-theory to the *cG*-theory. This asymmetry can easily be removed by placing Fig. 2 into the three-dimensional "space of the theories" in the *cGħ* frame of reference as was done by A. Zelmanov [186] (Fig. 3).

Bronstein's map had a history of its own. It started with Gamow, Ivanenko and Landau's contribution "The Universal Constants and the Passage to the Limit" that appeared in 1928 [156].

They started with a seemingly methodical problem of setting up a system of units and wrote that there were two methods of establishing a unit of measurement for a new physical quantity. It could be set up arbitrarily or, by relating it to the already established law that contained a numerical coefficient, one could select a standard unit that would turn the coefficient into 1. In the former case a new universal constant results; in the latter the number of fundamental (arbitrary) standard units and the universal constants remains the same. "As a result all we get is a natural unit (as compared to all the previous ones) that would measure our quantity".

One can also employ the second method to cut the number of universal constants by making one of them equal to 1. The co-authors called this reduction. They further wrote that "in the history of physics new universal constants and reduction of their number were manifested as sequence of theories and their gradual unification ".

To achieve complete reduction, one has to employ as many universal constants as the number of basic units in any given system. An abundance of physical constants and the fact that the *LTM*-system of dimensions is more popular than the others pose the question of which three of the constants should be selected. The authors suggested that "two heuristic suppositions" provided a key to this problem: the degree of the theory's generality represented by the given constant and a test for the constant's ability for the passage to the limit in the sequence: "classical theory – "vulgar" (semiclassical) theory – complete theory".

They admitted as "true" constants $\hbar$, $c^{-1}$ and $G$ and wrote that together with Planck one might arrive at dimensionless physics by obtaining "natural" units for all physical magnitudes.[5]

This led to their only practical conclusion (though related to the problem that was seen as important at that time): "without an electron theory one can draw a conclusion about its nature based on the dimensionality analysis": since $[e] = ([\hbar][c])^{1/2}$, $[m] = ([\hbar][c]/[G])^{1/2}$ then "the frequent attempts to fit a theory of a non-quantum electron into General Relativity are doomed to failure": if $\hbar = 0$, $c \neq \infty$, $G \neq 0$, then $e = 0$ and $m = 0$. They were obviously taking aim at Einstein and other enthusiasts of unified field theory who aspired to achieve $\hbar$-effects from the $cG$-theory that would be more general than General Relativity.

This was certainly a strange contribution (with not a single derivative or integral). It was destined to remain the three musketeers' only joint paper, and none of them ever returned to these ideas in later works. The least that Landau could have said about this was "philology".

Being prone to popular science, Gamow alone discussed these constants. His works, however, proved that the paper of 1928 left no trace in his mind. He voiced doubts about the fundamental nature of $G$ by suggesting that it was a camouflaged square of the weak interaction constant or admitted that $G$ was a variable rather than a constant or just denied that it was "a true constant" [153, p. 157]. Even when all three constants met in his works they were mentioned separately, rather than within the framework of $cG\hbar$ physics. In his *Mr.Tompkins in Wonderland* he used $c$, $G$ and $\hbar$ as capital letters for Mr. Tompkins' initials, but he easily changed the values of universal constants, in order to help readers better understand physics [149].

Bohr found this silly rather than funny [241, p. 189]; one may doubt whether Einstein would have found these "teaching methods" adequate. Both were convinced that the universal constants' values were indispensable for the world structure and that they could not be altered without deep changes in (or even destruction of) the entire edifice [310]. (This concerns the dimensionless combinations of constants.)

Bronstein shared this conviction and was very consistent in and insistent on his $cG\hbar$ viewpoint. It is hard to believe that he had no finger in the 1928 musketeers' pie. His unflagging interest in the correlation of theories and the key words "limits of the theory's applicability" that he used in the very first work in 1925 suggest another scenario.

In 1927 (the paper was dated November 1927) the "three musketeers", Jonny, Dymus and Dau, had just graduated from the university while the Abbot continued his studies. It was the Jazz-band's heyday; its members were full of *joie de vivre*. The paper was written in a restaurant to mark the birthday of a "fair lady" of the Jazz-band. The dedication was omitted by the editors yet it is more than clear that on the whole the paper did not fit the authors' own self-imposed high standards of scientific writing. The subject rather belonged to the oral genre of physics, which was "promoted" for the occasion.

This contribution pursued also another goal: its authors were far from being greatly impressed by the level of physics published in the *Zhurnal Russkogo Fiziko-Khimicheskogo Obshchestva* where Ioffe was editor-in-chief. In fact, they believed that the editorial board accepted contributions indiscriminately. It was their intention to demonstrate this with what they could call a "philological" article. *Physikalische Dummheiten* of which the authors were the editors carried the following equation on its cover:

$$ZhRFKO = \lim_{\hbar \to 0,\, c \to \infty} Phys.Dumm.^{6}$$

Probably it was Bronstein who had introduced the subject into Jazz-band talks; he might have been absent from the memorable dinner at which the article was written.

There is additional evidence – though being meticulous in his references Bronstein never mentioned this particular paper.[7]

### *5.2.3. At the Sources of Quantum Relativistic Astrophysics*

Bronstein looked for the signs of $cG\hbar$ physics not only in general considerations dressed up like a $cG\hbar$ diagram. He performed calculations. Some of them can be found in his works on the relativistic astrophysics.

In his article of 1933 [20] on white dwarf theory, he offered a detailed and clear physical discussion of the equilibrium of the gravitating sphere formed by degenerate Fermi gas for the non- and ultra-relativistic cases.[8] He was the first to obtain an equation for such a star for the general case when the degree of relativism was changing from the star's center to its surface [20, p. 99]. He wrote that this equation could be solved through "tiresome calculations". In 1935 Chandrasekhar made these calculations [296]; he advanced white dwarf theory to quantitative results that he obtained through numerical integration. He also wrote that he had supplied the equation in his preliminary note of 1934 [295]; it was known among the Soviet astrophysicists that it was Bronstein who first arrived at it [92].

However Bronstein's article, just like Landau's earlier work [214], did not deal with white dwarfs; what is more, they were never mentioned in them. The very titles indicated that the authors posed themselves a wider task – that of probing the stars' physical nature. Stars for them were mysterious *physical* objects; this partly explains why Bronstein never troubled himself with the concrete solution of his equation – he was not a "pure" astrophysicist.

In [20] he began by criticizing Eddington, who tried to describe the stars' structure without looking at the source of their energy. Then, following Landau, he discussed a gaseous sphere at zero temperature without an energy source. Composed of a classical ideal gas such a star cannot retain its equilibrium. It will contract until the quantum statistics become dominant. In this way Bronstein took as his subject the equilibrium of a sphere of degenerate Fermi gas. It should be noted that at that time it was generally recognized that the result of Landau's paper was not the critical mass for this configuration

$$M_0 \approx (\hbar c/G)^{3/2} m_p^{-2} \approx M_\odot$$

Bronstein wrote that E.C.Stoner was the first to obtain this remarkable result in 1930 [266]; however, Stoner failed to detect any problems in the unlimited contraction of the star with a mass bigger than critical; he believed that heating and radiation would be the only results.

Landau concluded that "all the stars heavier than 1.5 $M_0$ have a domain in which the laws of quantum mechanics (and hence quantum statistics) are violated" because with the mass more than $M_0$ "there is not a force in quantum theory that would prevent the system from contracting into a point"; on the other hand, he continued, "such masses are quietly existing as stars". He surmised that "all stars have this pathological domain" and that "they alone are responsible for stars being stars". He further suggested that "in the wake of Prof. Niels Bohr's beautiful theory" radiation of the stars could be "simply" suggested by the non-conservation of energy in relativistic quantum mechanics: this theory could be effective in the pathological domain when all the atomic nucleus compress in close contact and form one giant nucleus".[9]

As we can see, in astrophysics Landau progressed only as far as the $c\hbar$ theory; he regarded $G$ as an outside factor, the walls of a vessel, so to speak. Bronstein discerned in the astrophysical material an urgent demand for a $cG\hbar$ theory: "The relativistic theory of quanta that is a link between wave mechanics and special relativity should be further extended in accord with General Relativity" [20, p. 102]. Bronstein choose a simple example to support his thought: if the Sun contracts to reach nuclear density, its radius will be comparable with its gravitational radius.

He saw the necessity of creating a new physical theory applicable to any part of the Universe and wrote that "the very concepts of space and time and, consequently, the wording of the General Relativity principle and the gravitational equations will be radically changed to fit this new theory" [20, p. 103]. These underpinnings of the $cG\hbar$ diagram found their specific form in his doctoral thesis.

## 5.3. The Quantum Theory of the Weak Gravitational Field

Bronstein summarized his $cG\hbar$-study in two papers – a shorter one dated August 1935 ("The Quantum Theory of the Weak Gravitational Fields") and a more detailed article dated December 14, 1935 ("Quantizing Gravitational Waves").

The title and the length of the latter suggests that Bronstein made his thesis public. It comprised three parts. The first dealt with gravitational waves within the classical framework and introduced the next two parts, which discussed the quantum theory of the weak gravitational field in empty space and together with matter.

He proceeded from the general scheme of field quantizing suggested by Heisenberg and Pauli when he probed into gravitation in the approximation of the weak field that ignored the field's geometrical nature and treated it as a tensor field in flat space-time.

Just as Bronstein had promised, the paper burst with mathematical computations even though he omitted some intermediate stages – this was quite natural for the problem under study.

In General Relativity the weak field approximation is both rather special and sufficiently general – it has the maximum number of degrees of freedom while it is hard to separate coordinate effects from physical effects.

Let's look at the most specific features of his paper without going into technical details.

In the classical part Bronstein described gravitational waves using the four-index Riemann-Christoffel tensor (and not by a small additive term to Minkowski's metric as was the usual procedure). This allowed Bronstein to exclude all sham (coordinate) gravitational waves; he brought forth with all clarity the system's gauge freedom and the fact that the gravitational wave had two degrees of freedom (Bronstein used the terms *Eichungtransformation* and *Eichung*).

He obtained two significant results in the quantum part: he calculated the intensity of energy radiation caused by the emission of gravitational quanta and demonstrated that in the classical ($\hbar \to 0$) limit the quantum and classical theories of gravitation yielded similar results: Bronstein's quantum formula transformed itself into Einstein's quadrupole formula.

He applied to gravitation an idea voiced by Dirac and developed in 1932 by Fock and Podolsky, who arrived at the Coulomb force from quantum electrodynamics [281]. In the same way Bronstein reached the Newtonian law of gravitation as a consequence of the quantum gravity. He pointed out that the similarity of the two expressions notwithstanding the opposite signs of the forces naturally stemmed from the general quantum formalization.[10]

One could treat Bronstein's results as an obvious consequence of the correspondence principle and, at best, as evidence of the correctness of the applied procedure. In fact, the results were fundamentally important: the gravitation's special status and its association with the space-time geometry made the necessity of a synthesis of quantum theory and GR doubtful. It was generally accepted that gravitation was a special phenomenon far removed from the other physical fields; on the other hand, Einstein believed that GR was closer to a genuine complete physical theory than quantum theory. Bronstein demonstrated that the classical and the quantum (albeit incomplete) descriptions of gravitation were intimately connected. This naturally prompted a conclusion that a quantum generalization of GR was both indispensable and possible.

Bronstein never used the term "graviton" although it had already been coined by that time. It seemed to appear for the first time in print in an article by Blokhintsev [111], already mentioned in Chapter 4. By the time it was written, the theory of beta-decay based on the neutrino had been accepted. The last section, "The Nature of the Neutrino", offered some ideas related to the $cG\hbar$ story. Let's look at a longish quotation from it:

> Today the interaction of charged particles (Coulomb's law) [reference to Dirac's article of 1932] is regarded *dynamically*, namely, as the result of continuous emission and absorption of light quanta by interacting particles ...
>
> It is very interesting to compare the properties of the neutrino and the so-called *graviton*. So far, all the fields known to physics have been divided into two classes – *electromagnetic* and *gravitational* ...
>
> The numerous attempts – starting with a man of genius Faraday, and Einstein himself – to trace any connection between electromagnetic and gravitational phenomena inevitably perished in the jungles of formalism ...
>
> The radiation of electromagnetic waves is by far the only reason for the atom's instability. It also sheds its energy due to the radiation of gravitational waves by moving electrons of the atom, that are like planets in the solar system. Therefore, to penetrate the secrets of the atom's stability one has to surmise that the gravitational energy, just like the electromagnetic energy, are radiated by energy *quanta* rather than by energy waves – in the former case by quanta of electromagnetic energy (light quanta – photons), in the latter – by the quanta of gravitation energy (*gravitons*).

> They were of no consequence in the contemporary quantum theory of the atom because the calculated probability of their radiation is negligible as compared to light quanta in the same way as gravitational interactions are negligible as compared to electromagnetic interactions. It seems that the radiation and absorption of gravitons should have led to their interaction in accordance with Newton's law (to the gravitational field) in the same way as the absorption and emission of light quanta by charged particles lead to Coulomb's law. Just like light quanta, gravitons should have mass only when moving with the velocity of light. They are not electrically charged. From this point of view they are very *close* to the neutrino introduced by Fermi. The fact that the probability of graviton radiation is insignificant as compared to photon radiation is of consequence for the charged particles only. Being an uncharged particle, the neutron cannot radiate the light quantum therefore graviton radiation is very important for it. Beta-decay is one of the processes in which we deal with the *neutron*'s quantum transition. The comparison displayed above indicates that the graviton and the neutrino have much in common. This probably testifies that in general the highly improbable process of graviton radiation becomes practically observable in beta-decay. If the neutrino turned out to be a graviton this would have meant that contemporary physics had approached the limits beyond which there would be no present insurmountable barrier between gravitation and electromagnetism. Due to theoretical considerations it is hard to identify gravitons with the neutrino since it is hard to admit that they have the same spin 1/2 as the neutrino. In this respect gravitons have much more in common with light quanta. It is impossible, however, to totally rule out a theoretical possibility of their identification. So far it is much more correct to regard the neutrino as an independent type of particle.

All of this looks strange indeed. On the one hand, the authors seem to forestall Bronstein's conclusions; on the other, some superficiality, inaccuracies and lack of inner consistency are obvious. One feels that the authors contradict themselves.

Did they had priority over Bronstein? First, it was a popular article that described the situation rather than established priorities through careful quotations and references (they never referred to Einstein's famous statement of 1916). Second, at that time they were not concerned with the $cG\hbar$ physics – Blokhintsev's interest was the physics of solids [109]. And third, he was Tamm's student and collaborator, therefore it is hard to imagine Tamm keeping silent about his pupil's priority if it existed. What was more, Blokhintsev did not mention gravitation in their "Atomistics in Contemporary Physics" written in 1936 [112], which offered an ample opportunity to introduce gravitational quanta. Later, in the fifties, Blokhintsev showed skepticism about ties between particle physics and gravitation.

It is easier to surmise that the article was an echo of oral discussions in which Bronstein took part. This means, in turn, that his thesis, completed in the summer of 1935, was based on earlier ideas.

The idea of connections between gravitation and the neutrino was not exotic: Gamow wrote that back in 1933 "Niels Bohr posed the question about the difference between the neutrino and the quanta of gravitational waves". Gamow himself believed that "the ties between neutrino and gravitation were an exciting theoretical possibility" [152, p. 143].[11]

The echo of this question can found in the book *Gravitation, Neutrino and the Universe* by John Wheeler [273], one of Bohr's pupils.

This idea was alien to Bronstein [41] – at that time it had no profound reasons. At least, such reasons were absent from the *cGħ* frame of reference, which Bronstein used to look at the fundamental physical theories. Today, there are no such reasons either – there are no grounds to believe gravitation to be closer to weak than to an electromagnetic interaction. The neutrino could find a place for itself only in the dotted *cGħ* rectangle in Bronstein's diagram; it remains the same in present-day physics, anticipating the great unification of interactions.

The significance of the results Bronstein obtained in quantizing weak gravity did not depend on the changing fortunes of the unified theory. Speaking at a session when Bronstein defended his doctorate, Fock said: "The approximation offered by M.P. is beyond doubt. Even if Einstein's theory proves erroneous, Bronstein's results will stand" [173, p. 319]. Indeed, the results are relevant because the problem of correlation between the fundamental theories of gravitation and the area of their applicability is and will remain relevant.

From time to time physics generalizes a fundamental theory. In the 17th century gravitation got the *G* theory, in 1915, the *cG* theory, and it is now expecting a complete and consistent *cGħ* theory to be formulated. In theoretical physics "generalization" has a special meaning that includes both generalization and specialization, flexibility of individual components and rigidity (unambiguousness) of the theory's structure as a whole. Generalization is applied to quantitative predictions in specific problems, while, when applied to the theory's conceptual structure, the principle of correspondence is far from straightforward. In fact, theoretical physics is a hierarchy of models, the relationships between which being much more complicated than a simple relation of the general and the special (the *G*-, *cG*- and *cGħ*-theories of gravitation are one such). The "outmoded" models are not discarded but are preserved within the physical arsenal.

Bronstein's results on quantum weak gravity will retain their relevance. Afterwards there was a lot of attempts to calculate the elementary particles' reactions in which gravitons were involved. No matter how correct these calculations, they were physically meaningless. The result should contain a dimensionless multiplier of the $10^{-40}$ type and cannot have a tangible magnitude in the conditions of the theory's applicability. The quantum-gravitational effects could be tangible and even fundamentally important only under great concentrations of energy that would compensate for $10^{-40}$ found in astrophysical and cosmological conditions, where linear quantum gravidynamics is not relevant.

## 5.4 "The Fundamental Distinction between Quantum Electrodynamics and the Quantum Theory of Gravitational Field". The Quantum-Gravitational Limits

The previous section has somewhat distorted the real picture, since it deals only "top to bottom shifts": $(cG\hbar) \to (cG)$ and $(cG\hbar) \to (G)$. In actual fact, Bronstein began his work with an analysis of the measurability of the gravitational field related to the "bottom to top" shifts: $(cG) \to (cG\hbar)$ and $(c\hbar) \to (cG\hbar)$. This led to the discovery of the quantum-gravitational limits that are especially interesting for theoretical physics today.

### *5.4.1. The Problem of $c\hbar$ Measurability*

The development of physics urged Bronstein to analyze the problem of measurability. Its history goes back to the uncertainty principle (1927) that put the $\hbar$-limits to the use of concepts inherited from classical physics. The uncertainty relation limited only *joint* measurability of some (conjugated) pairs of values, such as a coordinate and momentum

$$\Delta x \Delta p \geq \hbar.$$

Yet, one could choose the degree of exactness of each value individually.

As soon as the meaning of $\hbar$-limitations had been grasped, there emerged a question of the nature of quantum limitations with relativism taken into account, or $c\hbar$-limitations. Mental experiments (starting with Heisenberg's microscope) produced as exact results as could be wished only by by-passing $c$-theory. Besides, the electromagnetic field, the key physical object, had been relativistic even before relativity theory was formulated: Maxwell's equations contained the $c$ constant. Heisenberg himself [158, p. 41], Fock and Jordan [280] all discussed the limits to measurability, or the uncertainty relation, for the electromagnetic field.

The physical community was especially impressed by Landau and Peierls' investigation of $c\hbar$-limits performed in 1931 [221]. An analysis of mental experiments in the $c\hbar$-domain produced not only *paired* but also *individual* uncertainties of the values that described the particle and the field. According to Landau and Peierls, the "field in a point" concept proved to be indefinable. This gave them grounds to doubt the contemporary quantum theory of the electromagnetic field and to forecast "a correct relativistic quantum theory (that has not yet been formulated) with neither physical magnitudes nor measurements in the meaning of wave mechanics" [221, p. 69].

This went hand-in-hand with other eloquent indications that the physics of that time was fundamentally deficient (the ±-difficulty of Dirac's equation, Klein's paradox, and so on). In the early thirties there was a wide-spread conviction that the $c\hbar$-reconstruction of many concepts (including the space-time concept) was inevitable. The most fundamental argument in favor of this was an analysis of field measurability.

Landau and Peierls believed that in their work they developed Bohr's ideas and provided theoretical substantiation for his non-conservation of energy hypothesis. However, Bohr did not accept this bold conclusion; in 1933 he defused it.

For Bohr their tendency to use only point charges as test bodies was their weaker argument. This idealization was borrowed from atomic theory. Bohr emphasized that "to verify the apparatus of quantum electrodynamics we may only manipulate test bodies of finite sizes the charges of which are distributed inside them. Any statement that unambiguously follows from that apparatus concerns values of the field averaged by the finite regions of space-time" [121, p. 128]. Bohr tied this proposition to the idea that the quantum electrodynamic apparatus did not include any supposition on the atomic nature of electricity. If one is prepared to accept complete freedom in determining the charge of a test body the limits indicated by Landau and Peierls are indeed removed.

Bronstein became interested in the $c\hbar$-theory at a very early stage. There were considerations connected with the physical values' observability and measurability that played a great role. In his 1931 review of Dirac's book Bronstein pointed out that the author underestimated the quantum-relativistic problems. To make his point, he quoted Pauli who had said, not without malice, at the 1930 Odessa congress: "*Die Observable ist eine Grösse, die man nicht messen kann*". Bronstein transformed it into: "The uncertainty principle of common quantum mechanics is too certain for the relativistic theory of quanta".

Bronstein responded with a short note on the measurability in the $c\hbar$-domain [24] to the work by Bohr and Rosenfeld [121]. Bronstein's comment was twenty times shorter and much more to the point of the mental experiments than the sophisticated deliberations of his eminent colleagues . They used imagined springs and massive frames together with arbitrarily large charges in arbitrarily small volumes (something that could not be found in nature).

At the same time Bronstein found a clear form for Bohr's conclusion that the $c\hbar$-limits did not spell death for the field theory.

Here are somewhat simplified estimates of the electromagnetic field $E$ intensity through the changed momentum of the test body with charge $Q$ and mass $M$:

$$E = \frac{p(t+\Delta t) - p(t)}{Q\Delta t}$$

Two summands make up $\Delta E$ uncertainty. The first of them is created by the uncertainty of measuring momentum:

$$\Delta E_1 \approx \frac{\Delta p}{Q\Delta t} \approx \frac{c\Delta p}{Q\Delta X}.$$

The second is a "reverse" field with the source of a current which is a product of the test body's charge and its velocity. Its indeterminacy (the velocity of recoil), which corresponds to the test body's localization with the $\Delta x$ indeterminancy, is equal to

$$\Delta v_{\text{rec}} = \frac{\Delta p}{M} \approx \frac{\hbar}{M\Delta x}$$

and the "reverse" field

$$\Delta E_2 \approx \frac{i}{c\Delta x^2} = \frac{Q\Delta v}{Mc\Delta x^2} \approx \frac{Q\hbar}{Mc\Delta x^3}.$$

By introducing the densities of the test body's charge and mass $\rho = Q/\Delta x^3$, $\mu = M/\Delta x^3$, with the body's volume being $V \approx \Delta x^3$, and taking into account that $\Delta p \approx \hbar/\Delta x$, we obtain

$$\Delta E = \Delta E_1 + \Delta E_2 \approx \frac{c\hbar}{\rho\Delta x^5} + \frac{\rho\hbar}{\mu c\Delta x^3}.$$

If we let $\Delta x$ tend to zero and admit that $\rho$ and $\mu$ tend to infinity sufficiently rapidly and according to different laws, we can see that $\Delta E \to 0$ for $\Delta x \to 0$. This justifies the concept of an "electromagnetic field in a point".

Bohr specifically emphasized that the field's indeterminacy conditioned by the influence of the test charge itself could be made, contrary to what Landau and Peierls thought, arbitrarily small; Bronstein indicated that to provide the *maximally exact* measurements of the field we should not strive to the lower radiation response to the test body. The final conclusion remained intact, yet Bronstein emphasized that at some point in the future the theory's potentialities would have to be correlated with nature's potentialities: "In the future relativistic theory of quanta, an impossibility of arbitrarily exact measurements of the field will be connected with the fundamental atomism of matter, i.e., with the fundamental impossibility to increase the charge density $\rho$ infinitely".

In his 1934 contribution, Bronstein justified the $c\hbar$ limitations for the electromagnetic field's measurability; no wonder that a year later he analyzed the gravitational field's measurability.

### 5.4.2. *cGħ Measurability and General Relativity Quantum Limits*

Let's follow Bronstein's train of thought and "Let's mentally experiment a little bit" (the title of a section in [30], see Appendix 1). The metric tensor $g_{ik}$ in the approximation of weak gravitational field

$$g_{ik} = \delta_{\iota\kappa} + h_{ik},$$

where $\delta_{ik}$ is Minkowski's flat metrics and all the quantities are $h_{ik} < 1$. As Einstein demonstrated back in 1916, in this case all general non-linear equations in GR are reduced to linear equations (to an accuracy of higher order of $h_{ik}$):

$$\Box h_{ik} = \kappa(T_{ik} - \frac{1}{2}\delta_{\iota\kappa}T), \tag{1}$$

where $T_{ik}$ is the energy-momentum tensor and $\kappa = 16\pi G/c^2$.

Having constructed an adequate Hamiltonian of the gravitational field Bronstein then followed the general scheme of field quantization suggested by Heisenberg and Pauli in 1929 and identified all commutation relations.

Before switching to a quantum picture of a weak gravitational field, Bronstein first tackled the problem related to the synthesis of the quantum and gravitational ideas in a more general case (not only for a weak field). After a short discussion of commutation relations, he wrote

> One might think that, as in quantum electrodynamics, there is a completely consistent quantum-mechanical scheme that contains the quantities which individually can be measured with arbitrarily chosen accuracy but taken together cannot always be measured with arbitrary adopted accuracy .... To understand the nature of those physical circumstances that can devalue this statement, let us take the simplest example of measuring the value [00,1], that is, one of the Christoffel symbols [that plays the role of the gravitational field's intensity]. This quantity can be measured through a test body that is moving with a speed that is infinitesimally small compared to the velocity of light [31, p. 214].

In this approximation, if we consider the gravitational field also to be weak, then the equation of the geodesic for coordinate $x^1$

$$\frac{d^2x^1}{ds^2} + \Gamma^1_{ik}\frac{dx^i}{ds}\frac{dx^k}{ds} = 0$$

transforms into the equation

$$\frac{d^2x}{c^2dt^2} = \Gamma_{1,00} = \frac{\partial h_{01}}{c\partial t} - \frac{1}{2}\frac{\partial h_{00}}{\partial x}; \tag{2}$$

here $x \equiv x^1$ and $\Gamma_{1,00}$ is the modern designation for the Christoffel symbol [00,1].

To measure the value of $\Gamma_{1,00}$, volume $V$ and time $T$ average (according to Bohr–Rosenfeld only these measurements can be applied in quantum field theory), we should measure the component $p_x$ to be the momentum of the test body with volume $V$ at the beginning and end of the time period $T$, since in the approximation under consideration

$$\Gamma_{1,00} = \frac{d^2x}{c^2dt^2} = \frac{p_x(t+T) - p_x(t)}{c^2\rho VT},$$

where $\rho$ is the test body's density. Therefore, while the momentum measurement has uncertainty $\Delta p_x$, the uncertainty

$$\Delta\Gamma_{1,00} \approx \Delta p_x/c^2\rho VT. \tag{3}$$

The uncertainty of $p_x$ momentum is composed of two summands: the ordinary quantum mechanical

$$(\Delta p_x)_1 = \hbar/\Delta x$$

(where $\Delta x$ is an uncertainty in the coordinate) and a "term connected with a gravitational field created by the measuring device itself due to the recoil during the measurement procedure". Bronstein said the following about the second summand. Taking into account the approximation, he transformed equation (1)

$$\Box h_{01} = \kappa T_{01} = \kappa\rho v_x/c.$$

If time $\Delta t$ is used for each individual momentum measurement ($\Delta t < T$ is a necessary condition), then the uncertainty of quantity $h_{01}$ connected with the uncertainty of the recoil velocity $v_x \approx \Delta x/\Delta t$ is of the order

$$\Delta h_{01} \approx \kappa\rho\frac{\Delta x}{c\Delta t}(c\Delta t)^2 = \kappa c\rho\Delta x\Delta t,$$

and according to (2) the uncertainty of the gravitational field's intensity is

$$\Delta\Gamma_{1,00} \approx \kappa\rho\Delta x.$$

In this case the corresponding uncertainty of the momentum connected with the test body's gravitational field is of the order

$$(\Delta p_x)_2 \approx c^2\,\rho V\Delta\Gamma_{1,00}\cdot\Delta t = c^2\kappa\rho^2 V\Delta x\Delta t.$$

In this way the total uncertainty of the momentum is

$$\Delta p_x = (\Delta p_x)_1 + (\Delta p_x)_2 \approx \hbar/\Delta x + c^2\kappa\rho^2 V\Delta x\Delta t. \tag{4}$$

As follows from (4), to minimize the uncertainty, we should choose

$$\Delta x = (\hbar/\kappa c^2\rho^2 V\Delta t)^{1/2}. \tag{5}$$

Then

$$(\Delta p_x)_{\min} \approx (\hbar\kappa c^2\rho^2 V\Delta t)^{1/2},$$

$$(\Delta\Gamma_{1,00})_{\min} \approx \frac{1}{c^2 T}\left(\frac{\hbar\kappa c^2\Delta t}{V}\right)^{1/2}. \tag{6}$$

Two conditions limit from below the time of momentum measurement $\Delta t$. First, it should be $\Delta t > \Delta x/c$ so that the recoil velocity caused by the changing momentum is lower than the velocity of light. From this and from (5), it follows that

$$\Delta t > \tau_1 = (\hbar/\kappa c^4\rho^2 V)^{1/3} \tag{7}$$

Second, from the very meaning of measuring the field in volume $V$ it follows that the value of $\Delta x$ should be smaller than the test body's dimensions: $\Delta x < V^{1/3}$. Taking into account (5), we obtain

$$\Delta t > \tau_2 = \hbar/c^2\kappa\rho^2 V^{5/3}. \tag{8}$$

Having obtained these two lower limits for $\Delta t$, Bronstein noted that the ratio

$$\frac{\tau_1}{\tau_2} = \left(\frac{c\kappa\rho^2V^2}{\hbar}\right)^{2/3} \equiv \left(\frac{c\kappa m^2}{\hbar}\right)^{2/3} \tag{9}$$

"depends on the test body's mass being infinitesimal for the electron and approaching the order of 1 in the case of a speck of dust that weighs a hundredth part of a milligram".

Correspondingly, the uncertainty $\Delta\Gamma_{1,00}$ thus obtains two limits:

$$(\Delta\Gamma_{1,00})_{\min 1} > \frac{1}{c^2T}\left(\frac{\hbar^2\kappa c}{\rho V^2}\right)^{1/3}; \tag{10}$$

$$(\Delta\Gamma_{1,00})_{\min 2} > \frac{\hbar}{c^2T\rho V^{4/3}}. \tag{11}$$

From this it follows that to make the maximally exact measurement of $\Gamma_{1,00}$ in the given volume $V$ we should employ the test body of the largest mass (density) possible; this makes the first limit the only significant one.

Bronstein pointed out that these arguments were analogous to the corresponding arguments in quantum electrodynamics. (He referred to his paper of 1934.) He wrote, further,

> Here we should take into account a circumstance that reveals the fundamental distinction between quantum electrodynamics and the quantum theory of the gravitational field. Formal quantum electrodynamics that ignores the structure of the elementary charge does not, in principle, limit the density of $\rho$. When it is large enough we can measure the electrical field's components with some accuracy. It seems, that nature knows of fundamental limit of the electrical charge's density (not more than one elementary charge for the volume with the linear dimensions of the order of the classical radius of the electron). Quantum electrodynamics, however, does not take them into account. ... Things are different in the quantum theory of the gravitational field: it should take into account the limitations caused by the fact that the test body's gravitational radius $(\kappa\rho V)$ should not exceed its real linear dimensions
>
> $$(\kappa\rho V) < V^{1/3}. \tag{12}$$

When this is taken into account (10) gives an "absolute uncertainty minimum",

$$\Delta\Gamma_{1,00} > \frac{1}{c^2T}\left(\frac{\hbar^2\kappa^2c}{V^{1/3}}\right)^{1/3}.$$

Naturally enough, "this was a rough approximation, because with the measuring device's relatively large mass departures from the superposition principle will, probably, be noticeable ..." However, Bronstein believed that "a similar result will survive in a more exact theory since it does not follow from the superposition principle. It merely corresponds to the fact that in General Relativity there cannot be bodies of an arbitrarily large mass with the given volume. Quantum electrodynamics knows no analogies to this fact – this explains why it can exist without inner contradictions". Having pointed out that this contradiction could not be removed from the theory of gravitation, Bronstein concluded:

> *The possibilities of measurement are even more restricted in General Relativity, with its arbitrarily large departures from Euclidean geometry, than can be concluded on the basis of the quantum-mechanical commutation relations. ... It is hardly possible to extend the quantum gravitational theory onto this sphere without revising classical concepts thoroughly* [30, p. 276].

This was how GR's applicability limits were first discovered. These limits were surmised – back in 1916 Einstein commented that quantum theory should have modified the theory of gravity; he voiced his dissatisfaction with the fact that the microscopic nature of the "rods and clocks" was disregarded; Klein also commented on it in 1927. All these comments, however, were of a logical or a methodological nature, while Bronstein made the analysis in the physical, quantitative language.

### *5.4.3. The Planck Scales in cGħ-physics*

Today the so-called Planck values are inevitably evoked in any discussion of GR's quantum limits. They are combinations of the fundamental constants $c$, $G$ and $\hbar$ of the type

$$A_{pl} = c^x G^y \hbar^z$$

and can have any dimensionality (of length, time, density, etc.). It is precisely the Planck units that are correlated with GR's limits resulted from the necessity of its quantum generalization.

The justifying arguments vary greatly – from the sketches of the future theory of quantum gravitation to dimensional analysis. Since dimensional analysis does not require any complicated calculations one can surmise that the quantum-gravitational role of the Planck units was obvious long ago, probably even to Planck himself [124].

In actual fact Planck suggested these units in 1899 when quantum theory had not yet been formulated. He suggested "natural unit measures" that would "preserve their significance for all time and across all cultures, including those outside mankind and the Earth" [254, p. 232]:

$$l_{pl} = (\hbar G/c^3)^{1/2} = 1{,}6 \cdot 10^{-33} \text{ cm},$$
$$m_{pl} = (\hbar c/G)^{1/2} = 2 \cdot 10^{-5} \text{ g},$$
$$t_{pl} = (\hbar G/c^5)^{1/2} = 5 \cdot 10^{-44} \text{ sec}, \qquad (13)$$

The so called Planck density

$$\rho_{pl} = c^5/\hbar G^2 = 5 \cdot 10^{93} \text{ g/cm}^3$$

characterizes the quantum-gravitational age in modern cosmology.

It was only in the mid-fifties that several physicists simultaneously pointed to the quantum-gravitational significance of these quantities. They were Landau, Klein, Pauli and Wheeler. (About the history of Planck's values, see [168].)

Implicitly, Bronstein used these units, since the three constants – $c$, $G$ and $\hbar$ – were involved in his analysis. It is very easy to complement his arguments in such a way that the Planck units become obvious – one of them (the Planck mass) is already present in Bronstein's writing: it is "the speck of dust that weighs a hundredth of a milligram" for which uncertainties (7) and (8) are of a similar order (the article [30] even has an explicit expression for the Planck mass).

To make the Planck values obvious, we may argue in the following way. We shall strive to measure the gravitational field not only with the least possible uncertainty but also in the smallest possible volume with the aim of determining the "field at a given point". In this case we should be forced to consider both limits (7) and (8) rather than (7) alone. To decrease uncertainty $\Gamma_{1,00}$ we should use the test body's maximally possible density; therefore (12) is

$$\rho = \kappa^{-1} V^{-2/3}.$$

In this case limits (7) and (8) are transformed into

$$\Delta t > \tau_1 = (c^{-4} \kappa \hbar V^{1/3})^{1/3} \qquad (7a)$$
$$\Delta t > \tau_2 = c^{-2} \kappa \hbar V^{-1/3} \qquad (8a)$$

By the very meaning of measuring intensity that is averaged through time period $T$, the condition $\Delta t < T$ should be satisfied; therefore, one should use the smallest possible $\Delta t$. Since $\tau_1$ becomes smaller as $V$ decreases, and $\tau_2$ increases, the minimal value of the largest of $\tau_1$, $\tau_2$ is achieved when $\tau_1 = \tau_2$. Then

$$\tau_1 = \tau_2 = (c^{-3} \kappa \hbar)^{1/2} = t_{pl}.$$

The corresponding dimensions of the test body

$$l = V^{1/3} = (c^{-3}\kappa\hbar)^{1/2} = l_{pl},$$

its mass

$$m = \rho V = (c^{-1}\kappa^{-1}\hbar)^{1/2} = m_{pl},$$

and, finally, the minimal uncertainty of the gravitational field intensity $\Gamma$

$$(\Delta\Gamma)_{min} = 1/cT.$$

If we take into account that the uncertainty in measuring the gravitational field should be assessed according to its combined effect on the test body – the work of the intensity for the distance of the order of the body's dimensions $\Delta g \equiv \Delta\Gamma \cdot V^{1/3}$ (the same quantity describes the metric's uncertainty), we shall obtain

$$\Delta g = l_{pl}/cT.$$

In this way the sphere of applicability of the classical theory of gravity-space-time is actually limited by the Planck values.

To reach the Planck range in quantum-gravitational phenomena, it is not necessary to analyze measurability (as Bronstein did in 1935) or to apply the Feynman integral (as Wheeler did in 1955); it is enough to introduce the $c$, $G$ and $\hbar$ constants. This can be done at the level of physics of 1913. Let us consider two point particles of mass $M$ tied together by gravitational interactions and moving along a circuit with radius $R$. Let's the system be rules by classical mechanics $Ma = Mv^2/R = GM^2/(2r)^2$ and Bohr's quantum postulate $2MvR = n\hbar$, $n = 1, 2 \ldots$. To find out which of the values of parameters $M$ and $R$ make the description of the system take into account the quantum relativistic effects, we should consider $n$ being close enough to 1 and speed $v$ to light velocity $c$. From this it follows that the closeness of $M$ and $R$ to Planck's values corresponds to the quantum-gravitational sphere. However, this leaves aside the profound space-time meaning of the $cG\hbar$-limits.

### *5.4.4. Reception of the Quantum-Gravitational Limits*

Even if Planck's description of the quantum-gravitational limits had been formulated by the thirties it would have taken some boldness to discuss them at that time – the magnitudes $10^{-33}$ *cm* and $10^{-5}$ *g* (= $10^{19}$ *GeV*) were worlds apart from what physics was operating with.

Here is what Heisenberg wrote in 1930:

The hope is ever-present that having solved the above problems [relativistic form of the quantum theory], the theory itself would be to a great extent reduced once more to the classical concepts. A cursory glance at the developments in physics in the last thirty years has convinced me that we should rather expect that the world of classical concepts would be further restricted. Our usual spatial-temporal world was changed when relativity theory claimed some of its characteristics were defined by the constant $c$; there are uncertainty relations in quantum theory symbolized by the Planck constant $\hbar$. With time they will be complemented by universal constants $e$, $\mu$ [electron mass] and $M$ [proton mass] [158, p. 79].

This was what physicists were thinking at the time. Any attempt to substitute $G$ for $e$, $\mu$, or $M$ would have been frowned upon: the ideas that could have connected (even in the most preliminary way) magnitudes $10^{-33}$ *cm* and $10^{-13}$ *cm* were conspicuously lacking at the time. (They appeared quite recently.) However, it followed from Bronstein's analysis and the general prospects of theoretical physics when viewed through the magic $cG\hbar$ cube (see Section 5.3) that there was some ground for such substitution.

Bronstein was aware of this when he wrote in [31]:

*To remove these logical contradictions* [connected with limited measurability of the quantum-geometrical quantities] *we should radically reconstruct the theory; in particular, we should renounce Riemannian geometry that is operating here with the quantities that could* [not][12]. *be observed in principle; it seems that we should probably reject the common ideas about space and time in favor of some much deeper and less visual concepts. Wer's nicht glaubt, bezahlt einen Taler.*

The pathos here is deliberately balanced by a phrase borrowed from Grimm's tale "The Brave Little Tailor".

This speaks volumes about Bronstein himself – his emotional intellect, wholesome self-irony and independent mind – who else would have dared to finish a learned contribution with a German proverb? It says a lot about the physical community of the time as well.

Theoreticians had already become accustomed to forecasts of space-time revolutions and were even a little bit bored with them – none had come true. Unlike the early thirties, in 1935 they were no longer restricted by great expectations: experimental discoveries of the neutron and positron and Fermi's neutrino theory opened wide vistas for the microphysics. It took uncommon courage to come up once more with predictions of the coming space-time revolution.

Decades later this conclusion was recognized, although the task of the quantum "radical reconstruction" of gravitation theory still remains a challenge.

Wheeler rediscovered the General Relativity quantum limits in 1955; two years later Planck's units returned from oblivion [167]. Today Planck's values are a common synonym of quantum gravity, the obvious limits of General Relativity. It is even more obvious that such limits do exist; back in the thirties there was no such conviction.

Bronstein treated seriously his analysis of measurability – he entered it practically in full in a concise exposition of his work. Nevertheless, the physical community failed to appreciate his conclusion that General Relativity was in principle restricted by quantum theory. In his review of Bronstein's [30, 31], Fock was very vague: "there are certain considerations of the field's measurability" [276]. While speaking at the defence procedure he praised the thesis as a whole [see also 275, 278] but voiced his doubts about the measurability analysis – for him the non-linear nature of General Relativity and of Born-Infeld electrodynamics seemed to be analogous; he likened the gravitational radius in GR to the electron radius in Born–Infeld theory [173, p. 318]. Bronstein was doubtful of such an analogy and proved to be closer to the truth.

There is a significant difference between the non-linear theory of gravitation (GR) and the non-linear electrodynamics of Born–Infeld. While GR's non-linearity is mainly unambiguous and the principle of equivalence is its physical cause, the non-linear lagrangian of Born–Infeld is "man-made":

$$L_{BI} = \frac{2}{\epsilon}\,(1 + \epsilon L_M)^{1/2},$$

where $L_{\mathrm{M}}$ is the usual Maxwellian lagrangian, and $\varepsilon$ is a smaller constant. This type of non-linearity did not follow from some profound physical principles; their theory, in fact, was tailored to deal with the electron's infinite proper energy while the lagrangian was specially constructed to prevent the problem from emerging on the classical level. The electron radius $r_{\mathrm{e}} = e^2/mc^2$ appeared to serve as a characteristic distance, the point at which the behavior of field $E \approx e/(r^4 + r_e^4)^{1/2}$ considerably differed from that of Coulomb.

Therefore one feels an urge to disagree with Fock, who insisted that there was an analogy between the electron radius in the Born–Infeld theory and the gravitational radius, the existence of which is conditioned by a fundamental physical fact – the equality of the gravitational and inertial mass (or the principle of equivalence as a theoretical expression of this fact).

Comparing mental experiments of measuring electromagnetic and gravitational fields, one can bring out the role of the equivalence principle. Let us have another look at the formula of the field uncertainty from Section 5.4.1.:

$$\Delta E \approx \frac{c\hbar}{\rho \Delta x^5} + \frac{\rho\hbar}{\mu c \Delta x^3}\,,$$

where $\rho$ and $\mu$ are the densities of the test body's charge and mass and $\Delta x$ is its size. This formula gives ground to the "electromagnetic field in a point" concept. This is due to the possibility of using the test body with arbitrarily large and *independent* densities of charge and mass. In physical terms this is fiction since so far we have been unaware of even mental processes of making such test bodies; this idealization, however, it admissible on the *theoretical* level, since electrodynamics does not contain any inner ban for it: the values for an elementary electrical charge and mass of the elementary particles are introduced from outside.

This ban does appear when we switch to gravitation; it can be formulated in different ways. It is clear, for example, that we cannot arbitrarily manage the quantities for density of the "gravitational charge" and the mass density since there is a connection

$$\rho_g = \sqrt{G}\mu.$$

It puts the equivalence principle, GR's cornerstone, into a nutshell. This is precisely what makes it impossible to decrease the gravitational field's uncertainty $\Delta\Gamma$ together with a decrease of the test body's dimensions $\Delta x \to 0$. Bronstein tied this to the impossibility of a test body with dimensions smaller than its gravitational radius.

It is even easier (though not convincing enough) to apply Bohr–Rosenfeld's remark to the theory of gravitation [121, p. 121]:

> All sorts of problems that appear in this connection can be discussed separately only because the quantum electromagnetic theory's apparatus is independent of the ideas of matter's atomic structure. This follows from the fact that only two universal constants – the quantum of action and light velocity – are parts of it, while two of them are obviously not enough to compose a characteristic length or an interval.

Relativistic quantum gravity needs *three* fundamental constants, $c$, $G$ and $\hbar$ that can make up a characteristic length $l_{Pl} = (\hbar G/c^3)^{1/2}$.

One should not be severe toward this analysis – a strict description of GR's applicability is possible only within the complete quantum theory of gravitation. Today, as in the past, the problem of the gravitational field measurability as well as the gravitational radiation of the electrons of the atom of which Einstein spoke in his time remain problems devoid of any practical significance. To analyze the theory's inner perfection, as Einstein used to say, the physicist should look at all possibilities allowed within the theory. Since the theory of gravitation does not ban deliberation of the atom and the quantum systems in general the range of the effect is irrelevant for an analysis of the theory by its own means. A mental experiment on gravitational measurability is nothing more than an instrument for probing the theory's inner perfection. The history of physics has repeatedly demonstrated that precisely the theories of inner perfection reached considerable outside justification by transforming technology and people's lives.[13]

Today, the problem of quantum gravity is no longer isolated – it will inevitably be a part of a unified theory of fundamental interactions. Back in the thirties Bronstein found the way leading to the very heart of contemporary physical research.

## 5.5 Physics and Cosmology

Bronstein's two last articles appeared in the March 1937 issue of ZhETF. One of them contained a nuclear-physical calculation made on Kurchatov's request. It could illustrate his everyday life in science; the second, "On a Possibility of the Spontaneous Splitting of Photons", reflected his own scientific interests. Related to quantum field theory (or to elementary particle physics, to use the contemporary term), it was relevant in cosmology. This was the first tangible result of marrying cosmology and elementary particle physics – something that is employed today when cosmological observations give a chance to reveal properties of elementary particles, while cosmological models are based on elementary particle theory. The situation was quite different in Bronstein's lifetime. Let's have a look at it.

### *5.5.1. Cosmology in the Thirties*

From the day of its birth in 1917 and well into the late twenties, relativistic cosmology was mainly regarded as an exciting possibility for the Universe's physico-mathematical picture and a proof of GR's immense potential and deepness. There were no verification means – experimentalists exerted themselves to verify GR's foundations and did not go too far. Cosmology's purely theoretical nature and numerous models of the Universe allowed a "loving condescension" one would extend to a promising yet too young a creature.

This changed radically in the late twenties. Astronomical observations, especially those by Hubble, had proved beyond doubt the out-Galaxy nature of the so-called spiral nebulae that turned out to be other galaxies. The intergalaxy distances fitted into the cosmological scale – this alone allowed them to speak of the Universe's homogeneity (outside this condition no more or less definite cosmological conclusions of GR's equations were possible).

The main result obtained by Hubble in 1929 through his studies of other galaxies was a consistent red shift in their spectra in proportion to distance

$$\Delta\lambda/\lambda \approx R. \qquad (14)$$

The only instrument that astronomy could employ to determine the speed of the distant objects was Doppler's interpretation of the spectrum shifts $v/c = \Delta\lambda/\lambda$ (if $v \ll c$). No special efforts were needed to formalize Hubble's correlation as a dependence of the galaxies' speed on the distance between them and the observer (our Galaxy)

$$V = HR. \tag{15}$$

Hubble obtained the coefficient $H$ through measurements and by introducing a multi-step range of distances: he calculated it as $H \approx (2 \cdot 10^9 \text{ years})^{-1}$. This was the first observational fact in cosmology. Very soon it turned out that relativistic cosmology had a mathematical model of its own that could describe this observational fact as well – it was Friedmann's non-static model of 1922, rediscovered by Lemaitre five years later[14].

This warmed up an interest in cosmology – it gave a possibility and, therefore, a necessity to verify qualitatively the cosmological theory through observation, that is, to treat it in earnest as a physical theory. Bronstein's relativistic cosmological survey [11] was one of the answers to the new situation. He concluded it by pointing out the central difficulty of relativistic cosmology connected with the Universe's age. The naive approach to the Hubble correlation (15) shows that expansion began

$$T = R/V = H^{-1} = 2 \cdot 10^9$$

years ago. The more exact cosmological model of Friedmann and Lemaitre made the Universe even younger. This contradicted the data of isotope geology (according to which the Earth was several billion years old) and astrophysics. It was thirty years later, after Hubble's scale had been revised several times that $H$ decreased ten times and, correspondingly, the age of the Universe became an acceptable $\sim 20 \cdot 10^9$ years. Today we are aware how poor cosmology's empirical possibilities were. Back in the thirties the astronomers realized that their observations could not produce exact facts; as happens, they lessened the degree of their deficiency.

The relativistic cosmology was tried hard not only because of the short scale of distances accepted at that time. The lack of correlation between observations and the theory bred alternative cosmological schemes. Some astrophysicists proved unable to accept the grandiose picture of the expanding Universe. They suggested simpler explanations of Hubble's correlation. In 1929, R. Zwikki suggested that the galaxies were not moving away – it was the photons that were reddening as they travelled to the Earth. In 1931, Milne initiated his "kinematic" cosmology, in which he ignored GR and which he sealed off from atomic physics. By contrast, Eddington made an attempt to build a fundamental theory that would describe everything as interconnected – both the Universe's micro- and macroscopic features. And these heretics were themselves the stars of first magnitude on the astrophysical sky. Physicists theoreticians mainly took this as sheer and groundless fantasies,[15] while the astronomers were diligently comparing all the models with observations. And GR cosmology didn't look observably attractive.

Not all physicists (even those well-versed in GR) were favorably inclined towards relativistic cosmology – its framework differed greatly from other physical problems. One has to identify in GR the description of a principally unique object with its principally unique history, while GR's mathematical apparatus (differential equations and initial conditions) does not differ from all other physical theories and cannot offer any uniqueness. A need for observational facts presupposed non-physical extrapolation since the Universe cannot be restricted to its visible part. Fock was one of those skeptical about such limitless extrapolation.

In the Soviet Union the predominant philosophy that occupied itself with the natural scientific developments among other things was weighing heavily on cosmology. While in physics the philosophical overseers limited themselves to separate consequences and interpretations (because physics' practical importance was too obvious), cosmology's modest achievements were a suitable target for attacks. Back in 1931 Bronstein's cosmological survey invited ridiculous philosophical commentary. By 1937, the time when his two last articles appeared, tension had grown. The self-appointed guardians of physics' philosophical purity found it hard to swallow the Universe's possible spatial finiteness and its non-static nature; they looked more favorably at the photons reddening. Ignorant in physics and dogmatic in philosophy they preferred to tread the well-known and officially approved paths. Naturally enough, notorious Lvov was among them. One of his numerous contributions to the struggle against idealism in physics was grandly titled "At the Cosmological Front". It was a powerful salvo that employed all permissible and non-permissible arguments short of a political interpretation concerning the photons turning red – in line with the arguments of another indefatigable fighter, Mitkevich, who attacked "physical idealism" deliberating the red meridian.

Some of these salvoes hit their target – witness what Landau wrote in one of his articles in 1937: "to maintain solar radiation on a stable level for two billion years (the surmised age of the Sun according to General Relativity theory) ..." [216]. Here he surely accepted Hubble's assessment of the Universe's age. But for these attacks, Zelmanov would not have published the second part of his survey "Cosmological Theories" [184] that enclosed relativistic cosmology.

This coincided with Bronstein's article that convincingly demonstrated that the hypothesis of photons reddening should be discarded. Bronstein was urged to write the article by an unexpected boost the hypothesis had gotten from fundamental physics (Dirac's concept of the electron vacuum).

In 1933 Halpern offered a hypothesis that in the vacuum the photon could interact with the negative-energy electrons. The result was their transfer into positive energy; by several steps they returned to the original while emitting quanta of lesser energy. Halpern wrote that "a scattering process of this type could only reduce the frequency; the reduction, if small, would on the average be proportional to the distance travelled by the quantum had to travel through 'vacuum'. In this connection the hypothesis may be mentioned that Hubble's constant might be reducible to atomic constants without utilizing gravitational theories" [142]. In 1936 Heitler, in his famous book, the first-ever monograph on quantum electrodynamics, supported Halpern [141, p. 193]. (It seems that the monograph was the direct cause of Bronstein's response, the brief version of which [34] appeared in 1936, see Appendix 2).

This support deserved detailed discussion. There are no grounds to believe that Bronstein was sure of a negative outcome – he never regarded the relativistic cosmology of his day to be final.

### *5.5.2. Bronstein's Attitude to Cosmology*

As early as 1931 Bronstein demonstrated his sober approach to a physical theory designed to describe the Universe no matter how much enthusiasm came through in his survey. At that time he believed that cosmology's main stumbling block would be the discrepancy between the Universe's age as determined by Hubble and the observational data. In later years Bronstein concentrated on relativistic cosmology's main theoretical difficulty, viz. the problem of choice created by diversity of cosmological models. It was his belief that a genuine cosmological theory should not only *select* a suitable model but also *explain* the fundamental properties of the Universe as a whole in an unambiguous way; it is its task, in particular, to clarify "why of two possibilities Nature preferred expansion to contraction" [21, p. 28] and to explain the dimensionless constants. It was his conviction that these problems could not be solved within non-quantum General Relativity theory, the equations of which were symmetrical in relation to the past and future. He wrote that a complete $cG\hbar$ theory was needed to solve these problems: "Today we have to demonstrate our ignorance when confronted with the most important questions about the world as a whole; physical theory has to traverse a long and thorny path before it would be able to raise and answer these questions in the proper way" [21, p. 30].

One would argue that within the last fifty years cosmology has posed and solved many problems (if they are not related to mathematics and cosmogony) and that Bronstein took a maximalist stand. At the same time, looking at cosmology today, one has to agree with Bronstein's statement. His thoughts contained a line that got extended in an interesting way into modern cosmology.

Einstein was not the first to speak about the Universe as a whole and of the extrapolation of physical laws onto it. Cosmology has inherited from classical thermodynamics and statistical physics the problem of the applicability limits of the second law of thermodynamics, "the Universe's thermal death" and the Boltzmann fluctuation hypothesis, designed to deliver the Universe of this dreary prospect. He suggested wc regard the visible World as a gigantic fluctuation in the equilibrium and on the average "dead" Universe; the probability of such fluctuations is low, yet they are inevitable. The hypothesis was heatedly discussed in its time even though the physico-mathematical language sometimes failed physicists.

Bronstein's conclusion was that Boltzmann's hypothesis should be rejected, since fluctuation probability was negligibly small and "since the Universe as a whole cannot be explained by laws symmetrical in relation to the past and future and substitution of $-t$ for $+t$. There should be at least separate domains in the Universe that are governed by the laws asymmetrical in relation to the past and future" [21, p. 14]. In 1933 it was easy to see that these "separate domains" could be correlated with the interstellar "pathological domains" [Landau, 1932] to which the *cħ* theory should be applied and where, according to Bohr, energy conservation was powerless. In his analysis, Bronstein did not depend on Bohr's idea – he was convinced that the Universe as a whole could be explained only by a theory asymmetrical in relation to time. His conclusion was that a complete *cGħ* theory would be such a theory.

Today, cosmology cannot do without $T$-non-symmetrical theories that explain, in particular, the Universe's baryon asymmetry [181].

So far Boltzmann's entropy theory has not been denounced through probabilistic considerations – they have never been shaped into a definite form that can be applied to relativistic cosmology. Yet Bronstein and Landau managed to introduce a new element to these discussions: they offered a clever argument in defence of the hypothesis and cleverly denounced it.

Bronstein wrote that in order to answer the question why man (or mankind) was lucky enough to observe huge-scale fluctuation, one had to point out that to emerge man needed the conditions impossible in the equilibrium, "dead" part of the Universe: "To observe any event we should exist; existence of life is connected with a whole range of conditions. It seems that the indispensable conditions include a firm planet heated by the Sun, etc., that is, a considerable deviation from thermodynamic equilibrium" [21, p. 13]. However, according to Bronstein, this qualitative answer fails a quantitative test: if fluctuation affected the domain stretching to the closest stars ($\sim 10^{40}$ km$^3$) rather than the entire visible Universe ($\sim 10^{66}$ km$^3$) man would have found enough space to survive. The relationship between these volumes describes the degree of the Boltzmann hypothesis' improbability.

The counter-argument lacks convincing force. For the observer's *existence* much smaller space would do for some time yet it can prove inadequate for *the emergence* of such an observer. Obviously, there is a gap between a possibility of fluctuation that would produce man and his environment "ready-made" and a possibility of the simplest fluctuation that would be the first link in the evolutionary chain leading to man. This gap can be described by a multiplier that is considerably smaller than $10^{-26}$. This shows that time $T$ should be long enough to allow such evolution, that is, there should be considerable space around the place where evolution is going on – $cT$. If the time of bio-evolution (~ $10^9$ years) that resulted in man is more or less correct, then the resulting domain is approximately the size of the visible Universe.

It takes no wisdom to see that the counter-argument is as vague as the original supposition, yet the initial condition exhibits a genetic connection between the object and the observer created by the object itself. This approach makes up the core of the so-called anthropic principle that invites much attention in cosmology today. It seems that this connection was first discussed (albeit in the negative sense) in Bronstein's 1933 article [21] and in [22], written together with Landau. Reproduced in *Statistical Physics* by Landau and Lifshits [219, p. 29] and in later editions it became the frame of reference for various "anthropic" speculations [185, 192].

Bronstein's attitude to the cosmological problem was mainly determined by his attitude to the fundamental physical theory since, he believed, "a cosmological theory is designed to crown the entire edifice of physical theory in general" [21, p. 29]. His views exposed in 1933–1935 were naturally branded with the maximalism of these years caused by general expectations of radical changes in fundamental physics. It seems that in later years his attitude to cosmology toned down a bit, yet the idea of the deep-rooted kinship between cosmology and microphysics was embedded in his mind.

Let's look inside the last scientific work of this mind.

### *5.5.3. Red shift, the Relativity Principle and Polarization of Vacuum*

Bronstein opened his work with a short discussion on the effects of non-linear self-action in electrodynamics. At that time the problem of light scattering on light was high on the agenda, both within classical non-linear electrodynamics (Born–Infeld) and quantum electrodynamics. Transmutation of the photon in empty space into several others, or the photon's spontaneous splitting, is a kindred process: if in the equation of scattering $\gamma_1 + \gamma_2 \to \gamma_3 + \gamma_4$ we move one photon from the left to the right of equation, the result will be the equation of photon splitting $\gamma \to \gamma_1 + \gamma_2 + \gamma_3$ Bronstein wrote on this account

> From the experimental point of view, the photon's spontaneous splitting is incomparably more interesting than light scattering on light. Today we are unable to verify experimentally the theoretical calculations related to such scattering – they require tremendous intensities in the conditions of a complete exclusion of any additional scattering. On the other hand, to observe the photon's spontaneous splitting (if it does take place) all we need is extensive time – no matter how small the probability of such splitting in a second is, it will occur if the photon is allowed to travel in empty space for a sufficiently long time. Astronomers have at their disposal the photons that were travelling in empty space for enormously long time (the light of the extragalactic nebulae of

> the 20th stellar magnitude reaches the terrestrial observer for 190 million years. Their spectra can still be photographed with the help of a 100-inch reflector.) Can this help to solve the problem of the photon's spontaneous splitting? [35, p. 285]

Recently, there has been much talk of using the Universe as a physical laboratory – there are both physical and economic limitations for setting up increasingly more powerful accelerators on the Earth [242]. Bronstein was aware of this possibility back in 1937. He wrote further:

> Halpern has voiced a hypothesis that the 'red shift' studied by Hubble and Humason could be explained by small infrared photons continuously flaking off the photon of visible light that reaches us from very distant celestial objects. Indeed, this is an attractive hypothesis, since all the red shift theories current today (the relativistic models of the expanding Universe and the diffusion of the galaxy systems according to Milne) failed to explain the observed quantitative value of the "expansion coefficient". At first glance one is inclined to think that the expansion coefficient of a galactic system can be deduced from the atomic physics constant with the help of Halpern's hypothesis. However, I think that Halpern's hypothesis is wrong.

Having revealed that "through the special relativity principle we can deduce some general properties of the phenomenon under study without referring to particular surmises about the nature of the spontaneous photon splitting mechanism", he strengthened his conviction with a elegant argument.

Let the photon is moving in empty space along the $x$-axis relative to a frame of reference $(x, t)$; in the Minkowski diagram $(x, ct)$ the photon's world line is $AB$. Another frame of reference $(x', t')$ moves relative to the original one with $v$ velocity also along the $x$-axis. If $w$ denotes the probability of the photon splitting in the unit

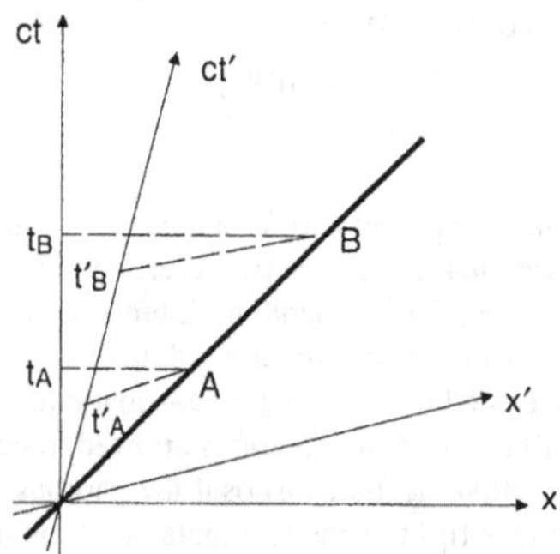


The world line of the photon

of time, then the probability of its splitting between two world points $A$ and $B$ should be the same in both frames of reference and be equal to

$$w \cdot (t_B - t_A) = w' \cdot (t_B' - t_A').$$

Applying the Lorentz transformation, we get

$$t_B' - t_A' = \frac{t_B - t_A - {}^{v}\!/\!{}_{c^2}\,(x_B - x_A)}{\sqrt{1 - (v/c)^2}}.$$

Since for light $x_B - x_A = c(t_B - t_A)$, then

$$t_B' - t_A' = (t_B - t_A)\frac{1 - v/c}{\sqrt{1 - (v/c^2}}.$$

From this it follows that

$$w = w'\frac{1 - v/c}{\sqrt{1 - (v/c)^2}}.$$

Since, according to the formula of the Doppler effect,

$$\nu' = \nu\frac{1 - v/c}{\sqrt{1 - (v/c)^2}},$$

then

$$w\nu = w'\nu',$$

that is, $w\nu$ is a Lorentz-invariant value. Bronstein demonstrated that $m\nu$ could not be considered a simple constant due to the uncertainty principle. If we take into account that

> the "photon" with which the experimenter is dealing is a quantum generalization not of the classical infinite flat wave but of a wave packet composed of such flat waves and, consequently, possessing not completely definite value of momentum

and if we introduce the parameters of the real "photon" – spectral width and indefiniteness of direction – with the help of the uncertainty of the wave vector $\Delta\nu_x$, $\Delta\nu_y$, $\Delta\nu_z$, we obtain

$$w = \frac{1}{\nu} f\left(\frac{\Delta\nu_x \Delta\nu_y \Delta\nu_z}{\nu}\right),$$

where $f$ is a certain function, that is, a strong dependence of the probability of photon splitting on frequency. In case this mechanism could explain red shift, the shift value would have to differ radically in different parts of the spectrum. According to astronomical observations, however, the shift is equal for all the spectrum lines of one and the same object, that is, in complete accord with Doppler's interpretation according to which the shift solely depended on the object's velocity:

$$\frac{\Delta\lambda}{\lambda} = \frac{\sqrt{1-(v/c)^2}}{1-v/c} - 1 = \frac{v}{c} + \frac{1}{2}\left(\frac{v}{c}\right)^2 + .$$

This killed off the cosmological component of Halpern's hypothesis.

The physical component could be verified solely through calculations – Bronstein made two such calculations although he wrote that "since the current positron theory is imperfect a complete theoretical solution of the problem is doubtful". The first of calculations made within the framework of the "elementary theory that took no account of the interaction between electrons of negative energy" resulted in the possibility of the photon's spontaneous splitting into three parts. It is hard to comprehend the resultant probabilistic expression physically. The second calculation took into account polarization of the vacuum and resulted in a zero probability of spontaneous splitting. This was the final sentence on Halpern's and Heitler's cosmological hypothesis.

The result based on the general physical principle – the relativity principle – was strengthened by direct quantum electrodynamic calculation; this was not easy at the time when (as Bronstein wrote) "there is no complete theory of vacuum polarization". Bronstein demonstrated his total command of fundamental physical principles, an ability to use them to obtain new profound physical propositions for complex situations, regard for experiment and, last but not least, skillful wielding of the theoretician's techniques.

The result was noted and appreciated. Zeldovich and Novikov cited his elegant argumentation involving the relativity principle in their definitive monograph on cosmology [183, p. 124]. It has also contributed to fundamental microphysics. This field sometimes attracts conjectures on the unusual phenomena at very small distances, that are small enough to be experimentally observable. Bronstein's argument points to considerable limits for conjectures of this sort. Indeed, if one conjectures that "something" may happen to the photon in the condition of very short waves when the discrete nature of space could be perceived, then according to Bronstein this "something" should always be aware of its low frequency tail that is easier to verify by experiment since

$$w \approx \frac{1}{\nu} \approx \lambda .$$

In 1937 Bronstein was not confined to the narrow limits of cosmology and the theory of vacuum (very narrow at that time no matter how strange nowadays we might think this to be). He wrote

> It is worth noting that the general properties of the probability of splitting in the unit of time derived above with the help of the relativity principle remain correct for the case of gravitational quanta into which the photon disintegrated together with light quanta. (These splits do not contradict the conservation laws.) Today there is no satisfactory theory that would explain the interaction between light and gravitation. Probably the future quantum "unified field theory" should consider these transmutations of the quanta of electromagnetic field into gravitational quanta whether complete or partial. "*Nature seems to be delighted with transmutations*". (Isaac Newton, *Optiks*, Question 30) [35, p. 290]

Transmutations into which the gravitons were involved attracted attention decades later within the limits of quantum gravidynamics. There is no doubt that they will be even more important in the "quantum unified field theory" of the future.

# Chapter 6
# Creative Personality

## 6.1. Perceiving the World

The reference to Newton quoted in Chapter 5 was not prompted by vanity – it reflected Bronstein's perception of the world. Physics for him was more than a game with the rules set once and for all. It was a construction that had been created by Newton and other natural scientists of the past. He was keenly aware that positrons and cosmology were organically linked with the Greek's atoms and Newton's mechanics. As a young man of 23 he was already aware of this connection:

> The world proved to be even simpler than the ancient Greeks imagined it: they believed all natural bodies to be made of four elements – earth, water, air and fire. Today we believe that protons and electrons are the basic particles that make up all material bodies. Will this conviction last? [63, p. 58]

He perceived the tie between the past and the future. In his *Atoms, Electrons and Nuclei*, speaking about the electron-positron pair's creation and annihilation, Bronstein quoted "a prophetic pronouncement" found in Newton's *Optiks*: "Nature seems to be delighted with transmutations. Why would it not transform bodies into light and light into bodies together with many other varied and numerous transmutations?" This prophesy, pronounced two hundred years ago, came true. Bronstein's quotation differed from the Russian translation of 1927; he read the English original. In 1929, when writing about atmospheric circulation, he referred to Galileo's *Dialogues* (a dialogue between Salviati and Simplicio on the causes of the passates).

Bronstein regarded physics as a humanitarian science because it was made by humans.

"Game" – this is how physicists often treat their research; this word has long been accepted as a term in its own right in psychology, pedagogy and cultural studies. In the same way as a child is easily carried away by his game, a physicist is fascinated with his manipulations of formulae and concepts. In his *Magister Ludi* Hermann Hesse invented a country in which everyone had mastering spiritual culture as his ultimate aim. Game was their mode of existence in which all the means would do – from Japanese verse and Bach to astrophysics and the theory of numbers. Writer made the distance between the exact sciences and the humanities look too short; he set his novel in the distant future yet he failed to put meat on the bones.

It was a physicist who said: "Physicists always want to explain complex things in simpler terms while poets are striving to make simple things look complex".

Bronstein would have hardly found this joke to the point. He would not have juxtaposed poetry and physics but would rather find them mutually complementary.[1] The physicist also divided sciences into the natural, the unnatural and the supernatural. Bronstein would have surely been defending the unnatural sciences, that is, the humanities. Common memory ascribes this classification to Landau, who used the word "philology" to condemn somebody's inadequate effort in physics. Though he treated the humanities with much more respect, Bronstein could have used the term disparagingly in the heat of discussion.

It was quite common among his friends to know a lot of poetry and to recite it quite often, yet Bronstein kept poetry in his heart rather than in his head. He brought a book of Pushkin to the Kharkov conference of 1934; the off-prints of his articles that he gave to Reiser bear poems by Schiller: "*Elisabeth / War deine erste Liebe; deine zweite / Sei Spanien!*" and O'Shaughnessy: "We are the music-makers / And we are the dreamers of dreams / Wandering by the lone sea-breakers / And sitting by desolate streams. / World-losers and world-forsakers / On whom the pale moon gleams / Yet we are the movers and shakers / Of the world for ever, it seems". (A slight misquotation shows that Bronstein wrote this from memory.)

Today not many physicists have a working command of several foreign tongues if only because English alone is sufficient. Back in the thirties more languages were used in physics – Matvei Bronstein stood heads above in this respect as well. He had a good command of three main European languages and acted as editor for translations from these tongues. He was very impressive when he alternated them at conferences. Born in Ukraine, he loved the Ukrainian language; he could write verse in Latin and delighted in learning Georgian, Spanish,[2] Hebrew, Turkish and Japanese. Obviously, he was confident enough to wander away from the well-beaten path of the Indo-European group.

In a way, his linguistic interests accounted for his fascination with the language of science – its ever changing vocabulary, semantics, idioms and the limits of expressiveness. Below we shall discuss his literary talent as reflected in his popular-science efforts.

Everybody who knew him agreed that Bronstein was an encyclopedically educated man who employed the amazing memory nature had given him with well-chosen and profound knowledge. He was always eager to learn. He was a passionate bibliophile and in book stores he discovered new lands for himself and his literary friends.

Here is what Frenkel, Fock [173], Mandelstam, Vavilov and Tamm [167] wrote about Bronstein:

> Matvei Bronstein is one of the most talented Soviet theoreticians of his generation. He couples an exceptionally wide knowledge of theoretical physics (ranging from solid body and atomic nucleus to relativity theory and the theory of quanta, statistics and electrodynamics) with splendid mathematical abilities;

> His powerful and critical intellect allows him to quickly probe the most difficult problems and makes him invaluable as a scientist;
> Matvei Bronstein is one of the outstanding Soviet physicists, endowed with rare erudition in many fields of physics.

His friends have never outgrew the impression Bronstein produced: "His knowledge was wide and varied. He was an erudite beyond comparison"; "Bronstein's knowledge was encyclopedic – he could maintain a professional discussion with anybody, be it a specialist in biology, ancient Egypt or paleontology, to say nothing of physics" [93, 238].

Here is what Kornei Chukovsky, a prominent Russian writer and cultural figure, had to say:

> Throughout my long life I met quite a few celebrities: Repin, Gorky, Valery Bryusov, Leonid Andreev, Stanislavsky. I know what admiration of an extraordinary personality is. It was the same feeling that gripped me each time I met the young physicist Matvei Bronstein. Half an hour was enough to realize that he was an uncommon man. With his inexhaustible knowledge of everything under the sun, he was brilliant company. English, Ancient Greek and French literature was as near and dear to him as Russian. You readily associated him with Pushkin's Mozart – the same animated charming and vigorous mind.[3] [167, p. 356]

He was one of those who did not regard physics as a means of beating others in the rush for completion – he was no pragmatist despite his powerful mind and complete mastery of mathematics. He refused to accept the shortness of human life as a pretext not to probe into problems that promised no speedy solutions – he was seeking an integral and developing physical picture of the world. His profound interest in the conservation laws and the limited nature of space-time concepts description in quantum-relativistic physics – two points that promised future growth – is an evidence of this.

This was much more than mere contemplation. He tied up these two problems with other fundamental facts and properties of physical reality: the sources of stellar energy, cosmological asymmetry, a genuine synthesis of the quantum and relativistic ideas that would become fact at some point in the future.

There is no doubt, however, that his attitude toward nature had a philosophical aspect as well. His popular science books and articles offer profound remarks of an epistemological nature, which were instrumental in mastering new physics [178].

His wide knowledge and powerful intellect made him a welcome participant in discussions. Yet sometimes they probably shackled his intuition. It seems that he himself was of the same opinion. This is what Frenkel wrote to his wife from the United States in 1931: "I was moved by Bronstein's letter in which he doubted his abilities and suggested that I get the Rockefeller grant for some other person. He has no ground to doubt his strength" [284, p. 267].

The commonly accepted truth that too much knowledge interferes with creative work in science cannot stand without crutches, that is, limitations. Indeed, there are people whom the burden of knowledge keep on the well-trodden paths or even paved roads. Others use knowledge to seek their own paths. There is every reason to believe that Bronstein was among the latter – he was never an over-zealous defender of obsolete foundations. We know that he was an enthusiastic fighter against some basic concepts.[4]

Speaking at a defence procedure, Bronstein remarked that the thesis to be defended strongly smacked of thc "university effect" and continued to explain that in mediaeval universities it was the aspirant's prime duty to protect the basic postulates. He was known to comment, not without malice, on certain experimenters: "They are afraid of making a great discovery". Does this speak of a man brimming with encyclopedic knowledge and afraid of a new edition of encyclopedia?

An idea that a theoretician's job consists in making discoveries is a superficial one – in a sense discoveries are but by-products. They depend on many circumstances: the general situation in science, the theoretician's preferences that might be called prejudices or a scientific ideal, his skill, his range of knowledge and other things that can be summed up as luck. Naturally enough, any discovery demands that a scientist should be completely engrossed in the problem – this depends on his perception of the world.

We have already written about theoreticians' varied perceptions of the world and identified two types of them – "thinker" and "pragmatic" or "doer". Although he could solve problems, Bronstein was never a "doer", in contrast to Landau, his university friend.[5] Close personal relations were doubled by shared scientific interests although their ideas of the world differed greatly. This probably explains why their friendship was never realized in more than one joint article.

Landau was past master of an art of formulating problems so as to make them solvable. Vitali Ginzburg put it this way: "The solution of complicated problems was Landau's strongest point" [163, p. 368].

E. Lifshits writes: "It was contrary to his nature to make simple things look complex – a trend fairly widespread nowadays (often justified by illusory generalization and strictness). Landau always moved in an opposite direction – to make complex things simple, to go down to the true and straightforward nature of things that underlie the laws of nature. This skill, his ability to "trivialize" things (as he himself put it) was Landau's special gift, and he was proud of it" [89, p. 14].

Indeed, this is an invaluable gift. It helped Einstein formulate special relativity based on simple kinematics instead of the complicated electron dynamics in ether.

However, a seemingly opposite gift – that of discerning complexity in trivial things (which later might lead to simplicity on a more profound level) – is no less valuable. General relativity, which made use of the commonly known fact of the equality of the inertial and gravitational masses to postulate curved space-time, is a good example of this.

In fact, the doer who can "trivialize" things is more successful, all other conditions being equal. The thinker is not as easily ignited – he is interested in the integral picture of the world. Naturally enough, different situations call for different methodological attitudes and ideas; on the whole, these two types complement each other.

Bronstein's life proved to be too short to allow him utilize his intellectual resources to the fullest extent; physicist-theorist's acme is over thirty.

Bronstein led an active life for which 24 hours in a day were definitely not enough; after all, his talent in physics was complemented by the talents of a teacher and a writer.

## 6.2. Vocation of a Teacher

Pedagogical talents flourished in Russia of the thirties: there was a shortage of teachers and university lecturers to train the swelling army of students; there was a lack of textbooks as well. This was especially true of physics, living through revolutionary changes.

It seems that Bronstein, who was a self-educated man, had many opportunities to think over teaching techniques. Though, naturally enough, the teaching talent could not stem from his self-educational experience alone.

He never begrudged his talent and was eager to address all kinds of audiences: senior pupils, students of physics, post-graduates and young academics. He was a successful author of popular science books for younger teens; on the other hand, he was also a first-rate speaker capable of presenting the most difficult problems. He was the main speaker at the nuclear seminar and spoke quite often at the theoretical seminar at his institute. He wrote quite a few popular science articles and books for adult readers – in short he was teaching physics to all who cared to learn.

It is hard to register the results of teaching (if not presented in book form) yet all who studied under Bronstein have preserved a memory of his talent and teaching skills.

Vonsovsky recalled that back in 1931 a group of graduates had invited Bronstein to teach a course on continuum mechanics at Leningrad University. He justified his reputation of brilliant lecturer (amazing in someone who had just left the university) and turned a boring subject into an adventure. The impressed students called Bronstein "Abbot Saint-Venant", that sounded like the name of one of the authors of elasticity theory. He used the breaks to tell his students about the physics that was developing before their own eyes. Great was their disappointment when Bronstein was replaced, although with famous Gamow (after a scandal with Hessen).

A. Migdal believed that together with V. Fock it was Bronstein who produced the strongest impression on him: "Brilliant in form and profound in content,

Bronstein's lectures taught us to love calculations – they were not rigid from the mathematical point of view (something that Fock liked very much) but perfectly adequate to the problem under study. It was a time when there were no textbooks on theoretical physics so his lectures were completely original. He was my first teacher of theoretical physics" [238, p. 23].

Here is how Y. Smorodinsky described Bronstein's lectures on electrodynamics (on the request of the book's authors):

> He started his lecture with the concept of a field. When he asked us where the light impulse's energy was when the impulse left the source but had not yet reached the detector we realized that the concept of a field was really very much needed. (Everybody knew that light was propagating with a finite velocity.) On the blackboard he drew a searchlight.
>
> He then explained that the field as a vector was affecting the charge. On the other hand, the source of the field, the charge's density, was a scalar. It turned out that the equation connecting the electrical field and density should be linear (the principle of superposition) and differential (the principle of locality). From this it immediately followed (the principle of symmetry) that $div\mathbf{E} = 4\pi\rho$ where $4\pi$ was introduced by tradition. Today this conclusion seems to be strict enough and all three principles in their proper places. At that time it challenged common sense. The result was incredible – we looked through all the textbooks and spent hours discussing the result and ... the first Maxwell equation became firmly rooted in our minds (yet there was a lurking suspicion that things were not that simple).
>
> Bronstein opened the next lecture with the law of conservation of charge. He pointed out that in order to satisfy $\rho + div\, \mathbf{j} = 0$ $div(\mathbf{E} + 4\pi\mathbf{j})$ should be equal to zero (taking into account the first equation. From this it followed (according to the tensor analysis rules) that $(\mathbf{E} + 4\pi\mathbf{j}) = c\ rot\mathbf{B}$, where **B** was a new arbitrary vector and $c$ a certain constant. Then Bronstein offered an unexpected conclusion: there should be another field besides field **E** – which was a magnetic field (consequence of tensor analysis rules!). Since magnetic fields had no sources (experiment!), $div\mathbf{B} = 0$. During the break the audience was boiling with indignation but failed to topple Bronstein's logic. In this way the students memorized two more equations.
>
> The latter was deduced from Faraday's law. Everything proved to be tied together. Discussion of specific problems followed – there were more effects though everything was strict enough. The students continued to boil and check the textbooks.
>
> Here is another episode. Bronstein started the lecture with the words: "Today every fool knows how to integrate. Let's try something really challenging – let's learn how to differentiate". He proceeded to prove that the solution formalized in the form of retarded potentials satisfied Lorentz's condition $d\varphi/dt + c\ div\ \mathbf{B} = 0$ due to the law of charge conservation.
>
> Derivatives and integrals change their places, limits of integration tended to infinity. We all felt that something of cosmic dimension was going on the blackboard. Again it produced an indelible impression on the students.
>
> His lectures were like an abstract theatre, paradoxical and grotesque. They remained with us as artistic creations.

Bronstein had vast teaching experience: he lectured at the university, at the department of physics and mechanics of Polytechnic and at the Teacher Training College on practically all fundamental physical courses: electrodynamics, statistical physics, quantum mechanics, the theory of radiation (the earlier name for quantum electrodynamics), the theory of gravitation and others. In the academic year 1934–

35 he taught a course on the theory of atomic nucleus (then quite a new branch) for young academics at the Physicotechnical Institute.

He selected the shortest path and was never tempted to embellish his lectures with historical anecdotes and amusing details. In fact, despite his young age and a preoccupation with his own research, he was well-versed in the history of science. Yet he was aware of scientific logic in historical background and of history as exposed in the bright light of logic. He was not only teaching physics but also how physics was done; he explained complex scientific constructs and demonstrated a new physical style.

Every age in physics has a style of its own. Einstein was the pioneer of new style in 20th-century physics. In the Soviet Union this style was formed in the thirties, Landau's fundamental course of theoretical physics being its best-known presentation. Bronstein was one of those who contributed to the new style.

It freed scientists from the pinched limits of induction that prescribed progress from lesser facts to more significant ones and through them to laws and further on to principles. Though experiment, as the theory's supreme judge, remained as important as ever, physicists made active use of symmetry, the most general formulation of problems and analysis methods. Formalization was everywhere though physics was not a substitute for mathematical formulae – the opposite was true: mathematical concepts were acquiring physical meaning. This style dominated 20th-century theoretical physics; it is only recently that there appeared the first signs of a still newer style.

Bronstein's lectures were marked by the approach best known from the Landau–Lifshits course. Both Bronstein and Landau were interested in the practical pedagogical problems they encountered in their teaching activities. They discussed the problem of the minimum amount of knowledge any theoretician would need; it seems that this gave rise to the idea of a definitive course on theoretical physics that would bind together a set of textbooks written by like-minded people and employing the same style and approach. In the early thirties Landau became inflamed with the idea of shaping the most advanced theoretical physics in the Soviet Union – the theoretical minimum and definitive course should have been his main instruments; hence his close cooperation with Bronstein, who was a gifted writer and popularizer.

The original idea was that different volumes would be written by different people: Landau asked Pyatigorsky, one of his post-graduate students, to write the first volume, on mechanics. He himself wrote the table of contents and meticulously edited the manuscript, making it as concise and precise as possible. Pyatigorsky acknowledged that the book owed its merits to his teacher. The first issues carried the words "Edited by L.D. Landau", probably an indication of the planned unified style and approach to volumes written by different people. The first volume, on mechanics, appeared after the second volume, on statistical physics, was published.

Bronstein undertook the task of writing a textbook on statistical physics, the absence of which was acutely felt. Distinct from earlier books on the same subject, Bronstein based on the Gibbs method, the most general method of statistical physics.

Y. Smorodinsky preserved material evidence of cooperation between Landau and Bronstein – three slim exercise books with the words "M.P. Bronstein and L. Landau. Statistical Physics" written on each of them. The poem "Death of a Poet" and a picture of "Pushkin's Duel" were reproduced on their covers. It was 1937, the centenary of the great Russian poet's death and a horrible year for Russian intelligentsia.

This is the story behind the three exercise books. In 1937 Smorodinsky, a third-year student at Leningrad University, asked Bronstein, who was his lecturer, for a theme for his annual project. After a short interview Bronstein advised Smorodinsky to read up on statistical physics and lent him a type-written manuscript.

The book "Statistical Physics" was sent to the printer's in October 1937 after Bronstein had been already arrested. It appeared in February 1938. Late in April 1938 Landau was arrested and kept in prison for a year.[6]

"Statistical Physics" was intended to be the second volume of the course that should precede "Quantum Mechanics" – this explains why quantum statistics was left out. Bronstein himself was quite prepared to treat this subject – volume two of the *Dictionary of Physics*, issued in 1937, should have contained his article on quantum statistics [42], only one copy of which has survived.

The last sentence demands clarification.

When leafing (for the *n*th time), in a central library, through the *Dictionary of Physics* (which appeared between 1936 and 1939) I looked on the final page of Volume 2, which carried the publisher's information. I knew that only the first (out of five volumes) contained Bronstein's articles. It had never seemed strange until I saw that Volume 2 had been sent to the printer's on March 30, 1937, that is, four months before Bronstein's arrest. Where then were his articles? Lost in thought, I turned the pages and to my amazement saw Bronstein's name at the end of the entry on quantum statistics. I realized that it had one beginning and two ends signed by two different people; the pages had the same numbers. Further examination revealed the truth – fifty years ago the volume had been doctored in the print-shop. By mistake a printer did not remove the page with Bronstein's name on it as he was expected to do. It seems I was holding a unique copy, one in thousand: clearly this was a mistake that could easily have cost the worker, if not his life then his freedom.

This gives us an opportunity for a glimpse of life in 1937, the last year of life for many outstanding people in Russia.

Here is an obedient young physicist writing an article to fit into the slot taken up by the previous article. He is not even writing the entire article – he is continuing it starting with half a word – the publisher saved some money.

Here is Vladimir Fock, the author of "Quantum Electrodynamics", an entry that came after Bronstein's, who discovered the substitution when he got the finished volume.

Here is Matvei Bronstein, who will never see it. He is kept in an overcrowded cell where people sleep on a cold cement floor. In the daytime he tells his cell-mates about Ancient Greece, the French revolution and astronomy. He never speaks about quantum statistics: nobody in the cell is interested in it ...

Let us go several years back to when Bronstein was lecturing on quantum statistics and other subjects at the university. He spoke to wider audiences as well. One gets an ample idea of his gift as a teacher and writer by reading his popular-science books and articles. The most important of them appeared in 1935.

His book *Atoms, Electrons, Nuclei*, intended for senior pupils, tells how ideas of atomism developed. With its obvious merits – simplicity of presentation, concise and profound information – it was the natural choice forty-five years later as the opening volume for the newly established "Library of the *Quantum*".

*Structure of Matter* was addressed to the more mature reader consciously seeking knowledge. The author himself stated that "the aim is to explain in a simple and clear form what contemporary physics knows about the structure of matter. This is not a complete or perfect theory – every year brings new discoveries that make radical changes in our ideas of the world inevitable. No book on physics today can or should claim a rigid picture of physical theory. It should rather try to show how it changes so that the readers can see the *direction* of these changes. This book aims precisely at that when looking at the path traced by the ideas of matter from Democritus and John Dalton to the latest discoveries in this field" [81, p. 3].

When reading the book one finds it hard to believe that this vast amount of knowledge could be squeezed into such a small volume. There are four chapters in it: "Atoms and Molecules", "Electrons and Nuclei", "Quanta", and "The Universe". Actually, this is a course of general physics with the center of gravity being shifted to modern physics. Bronstein gave much space to the theory of relativity, quantum mechanics, quantum chemistry and cosmology. All this was introduced by a discussion of classical mechanics and electromagnetism. This is a systematic course, the mathematical apparatus of which is limited to the four rules of arithmetic. No wonder it was used as a textbook in higher educational establishments. Here are some extracts that demonstrate how Bronstein presented novel and complicated ideas.

Having explained the relative nature of the idea of simultaneity he wrote:

> This may seem strange to those who treat the ideas of time and space metaphysically as predating any experience. They treat them as the glasses through which we have to look at nature irrespective of its real properties. In actual fact we are not in a position to separate the ideas of space and time from the material bodies that fill in nature. Therefore, the laws of space and time and even the very

> possibility of their use are but part of the general system of laws of behavior of material bodies. The laws cannot be guessed (i.e., known before experiments and research) so we should accept them and adjust our thinking habits accordingly. (It happens quite often that we are baffled by them due to the habits and prejudices we acquire under the influence of everyday experience, the sphere of which is immeasurably narrower than the sphere of scientific experience in general.)

He further pointed out that the similarity between electron mechanics and the laws of wave propagation was all but rather shallow. Bronstein explained: "In this connection it is becoming increasingly clear that the question so loved by popularizers – is the electron a particle or a wave? – is a sheer misunderstanding. The wave is a process, while the electron is a thing – obviously the electron cannot be a wave. On the other hand, the surmise that the electron is an elementary particle only means that we are unable to observe its smaller parts under any condition. In this respect this surmise is one hundred percent true. The correct answer to the notorious question is: the electron is a particle that behaves according to the laws of wave mechanics."

His discussion of the elementary nature of the proton and the neutron in connection with beta-decay led him to a conclusion: "There are relationships in nature that overflow the limits of our visual ideas about the whole consisting of parts." Later developments in physics have provided this conclusion with even more solid support when the quark nature of the hadrons became a fact.

## 6.3. Science and Literature

Bronstein published his last popular-science works in 1935; this should not be taken to mean that he abandoned writing in general – he began work in real literature. In his last two years he wrote for teens, which is by no means "lesser literature".

In the twenties and thirties literature seemed to be fascinated with the recent scientific developments – Platonov's hero was feeding electrons while Woland in Bulgakov's *The Master and Marguerita* was successfully using a five-dimensional theory. Science fiction was extremely popular. The novel *Time, Space, Movement* was about "the years of reaction of 1907–1911".

The number 1 Soviet writer of the time, Maxim Gorky, contributed a great effort to create a new society and wrote about the role scientific knowledge and literature could play in this process. He had educated himself through reading and was especially interested in developing children's literature. It was his conviction that academics and writers should cooperate in writing books for children. His article written in 1933 began with the words: "The question of subjects of books for children is, undoubtedly, a question of children's social education". He aspired to create "a new children's book" that would "popularize scientific knowledge in artistic form" and stated that "there should be no sharp demarcation between *belles-lettres* and popular-scientific books" that, he insisted, could be possible only if true scientists and the best writers would pool their efforts. He said that authors should be recruited from among scientists, not indifferent compilers. He finished his article with a push to discuss the project carefully and set up a group of young writers and scientists without delay.

In 1933 there were several meetings between Leningrad writers and academics in the Physicotechnical Institute. The *Literaturny Leningrad* newspaper mentioned writers M. Zochshenko, V. Kaverin, L. Leonov, S. Marshak, Y. Tynyanov, K. Chukovsky, physicists Y. Dorfman, A. Ioffe, N. Semenov, Y. Frenkel and mathematicians B. Delone and M. Frank. They discussed what was similar and dissimilar in both types of creative work, how one should write about science and depict scientists in works of fiction. Science had already entered everyday life through new technology so *belles-lettres* were trailing behind. They also discussed a joint almanac (much later realized in the popular series "Travels into the Unknown"). Y. Dorfman published his long story "The Magnet of Science" in a collection which Gorky patronized.

Bronstein came to children's literature through a different door – his road proved to be much shorter. His wife, Lydia Chukovskaya worked with Leningrad Children's Publishers. She edited all his books while Samuil Marshak was their "editior-in-chief" [298].

It should be said that the very genre of Russian artistic-scientific literature was born in Marshak's vicinity [248]. (In Russian, there is a term "artistic-scientific" literature to distinguish such books from popular science. An example of such literature in English are books by Paul de Kruif.)

Marshak later wrote: "We were captured by the idea that in children's literature fiction and knowledge could go hand-in-hand, they were not separated as in books for adults" [236, p. 171]. The main problem was to find authors who could combine these two elements – one had to be extremely lucky, much like a treasure-seeker. Indeed, such a person should be a scientist so that his story would be based on authentic material. On the other hand, he should be a writer to be able to smelt his life and scientific experience into an artistic book. Naturally enough such people were few and far between. The most amazing thing was that Marshak managed to find them.

Bronstein was one of these finds. Marshak was quite fascinated (as he was fascinated with all his finds) with the young physicist who, he felt, could write artistic books. He easily ignited people with his enthusiasm. Bronstein was carried away by the new challenge – to combine the logic of scientific thought, the logic of feelings and the logic of words into one logic, the artistic logic of a book. Being overburdened with work he spent much time writing and rewriting his first book under Marshak's supervision.

The first step proved to be an easy one – he selected the theme, spectral analysis, that allowed him to show science "not as the storage place of ready-made discoveries and innovations but as an arena of struggle upon which man is trying to soar despite the burden of traditions and resistance of his material" [174]. Marshak, however, was not satisfied with the draft of the very first chapters that came habitually easily to Bronstein. Together they altered it many times before they outlined the plot – the discovery of helium and the key words, "Solar matter". What was even more important, Bronstein developed his own approach to the text.

His earlier popular scientific works were brimming with bright metaphors, emotional presentations and patches of well-organized sentence structure. In some places, however, they were marred by complicated syntax and stretches of dull, if not boring, descriptions. Sure enough, the adult reader conscious of his desire to learn could surmount the difficulties which could prove too much for a teenager awakening to the frightening complexity of the world around him. One should be clear and precise when writing for teens – something that could come only as the result of hard and persistent efforts.

Indeed, every writer strives for exact words and genuine intonations, yet Marshak outdid them all – these were his faith. Being keenly aware of each author's specific style, he insisted that every word and comma should be assigned the right place, where nothing else would do. Thus it was for the first time that Bronstein was exposed to this sort of editing, yet he adjusted very quickly to the high requirements.

The talent of a writer was part of his personality. It was coupled with his talent as a lecturer, able to build up his material, find the best form for it and be keenly aware of the audience's response. This is what any writer needs as well.

On the other hand, the task that Bronstein was bold enough to shoulder was far from easy – after all, his lectures were attended by people who were consciously seeking knowledge. Schoolchildren, though inquisitive, knew too little to be treated in the same way. Writing for them was more difficult and easier at one and the same time – psychologically Bronstein was closer to children, less preoccupied with the everyday affairs in which adults are steeped. People like Bronstein preserve throughout their lives an interest in the world around them that is present in every child.

The first book demanded a lot of effort. Bronstein was always very conscientious in everything he did, and one could not work differently when dealing with Marshak and his team. Bronstein had read back issues of scientific and popular journals for half a year when doing spadework for his *Solar matter*. Its first edition, which appeared in 1934 in a children's magazine, differs greatly from its final version, published in 1936 under separate cover – a sure sign of meticulous polishing.

The result was brilliant – this was true literature. It is hard to imagine anybody would put the book aside after reading the very first sentence: "I shall tell you about a matter that people first discovered on the Sun and only later found it on Earth". The table of contents was no less tempting: "Color Signals", "A Failure", "Plain Glass", "Decoded Signals", "Ashes, Granite and Milk", etc.

Marshak was also satisfied; he even penned a foreword [235] that stated the program of artistic scientific books for children. He wrote that in the past ignorant mediocrities with no faith in a child's inquisitiveness and science's attractiveness were doing hack-work with all kinds of cheap tricks to keep a reader riveted to the book. He further stated:

> False and sly didactics is not what we are after. We respect science and have a great regard for our small readers. Being aware of their demands, we reject simplification and strive for a clear and consistent exposition. True enough, a child wants a book to be interesting, but this can be achieved by emotional presentation, varied and rich thoughts rather than by tricks alien to the book's subject. Only an enthusiastic author who has the problem at his fingertips is free and confident in his presentation.

Marshak continued that the author should know how to bypass excessive terms, an ability "given to those, who being exposed to the strictness of the language of science, have not lost the taste for the living tongue". Only this will allow us "to assess popular-science books for children with the yardstick designed for adult *belles-lettres*".

Marshak recalled later: "The very memory of our joint work with Bronstein is near and dear to me. Our cooperation with Dorfman, who was a journalist besides being a physicist, was a complete failure; our cooperation with Bronstein was a complete triumph. His work was much closer to artistic literature than Dorfman's publicist writing – with him it was a patchwork of sham *belles-letters* and perfect bores" [236, p. 173].

Kornei Chukovsky wrote:

> Being a children's writer myself, I can confirm that Bronstein's *The Solar matter* and *X-Rays* are superb. They are more than just another popular-science effort – they are elegant and artistic – in fact, they are just sort of poems about the greatness of man's genius. Their enthusiasm, a precious and rare quality, is catching. All newspapers and magazines were singing praises to their author. I felt happy that Soviet children had acquired another friend and tutor. I was doing my best to persuade him to continue writing because science popularizers are as rare as talented writers [167, p. 357].

In his Foreword to the 1959 edition of *The Solar matter*, Landau agreed with the writers who knew nothing about physics: "This is a clear and absorbing book equally interesting for the pupil and the physicist. Once you open it it is hard to put it aside".[7]

Practically all the central papers and journals responded to the book.

Georgy Adamov (a popular author of science fiction) published a detailed review [88]:

> This book – clear, light and absorbing – was written for children by a Soviet chemist. He is a happy blend of profound scientific knowledge and literary talent. One has to be well-versed in one's field to be able to extract confidently all the necessary facts from science's inexhaustible treasure-store. One has to be a gifted story-teller with a fine ear for words and syntax to be able to lay out the facts from the history of chemistry and physics in this attractive and inviting way.

Дорогой Лидочке, без которой я никогда не смог бы написать эту книгу.
Митя 21 апр. 1936

**М. БРОНШТЕЙН**

# СОЛНЕЧНОЕ ВЕЩЕСТВО

**ОБЛОЖКА И РИСУНКИ**
**Н. ЛАПШИНА**

**ЦК ВЛКСМ**
**ИЗДАТЕЛЬСТВО ДЕТСКОЙ ЛИТЕРАТУРЫ**
**ЛЕНИНГРАДСКОЕ ОТДЕЛЕНИЕ**
**1936**

Title page of *The Solar Matter* with dedication to L.K.Chukovskaya.

It is not as easy to learn what children thought about the book. We were lucky enough to meet one of the readers. He is Prof. V. Ivanov of Leningrad University, an astrophysicist who commented on Bronstein's early astrophysical works for this book. When discussing the Hopf-Bronstein formula obtained in 1929 he described his personal feelings toward to Bronstein.

In the late fifties, as a post-graduate student, he was studying the theory of radiation transfer. Among the many illustrious names in the bibliography attached to Hopf's classical book [294], he stumbled across the name of a certain M. Bronstein.

> The name rang a bell, yet I could not place it. Then I realized that Chandrasekhar, in his famous *Radiative Transfer* (1950) mentioned the Hopf-Bronstein formula. Still, I felt that I knew the name from some other source. Then it dawned on me that I had read his article in the collection *Problems of Cosmic Physics*, published in the mid-thirties, that was standing on my bookshelf. This meant that Bronstein was one of us. And then, like lightning, came the realization that it was Bronstein who had written that marvelous book *The Solar matter*! Immediately after the war, probably in 1947, I had borrowed it from the Leningrad Club of Scientists and never stopped reading until I finished it. I had already known that I would be an astronomer and was reading everything I could find about astronomy. So the Hopf–Bronstein relation (the term introduced by Chandrasekhar) belonged to *my* Bronstein who had decided my destiny.

*The Solar matter* was followed by two other books – *X-Rays* and *Inventors of the Radiotelegraph* – that needed practically no editing [298, p. 293]. It seems that Bronstein mastered the skill of being a children's writer fairly quickly.

It is interesting to note that for children's books Bronstein preferred subjects far removed from his professional interests: experimental physics, chemistry and technology (no wonder the reviewer took him for "a chemist"). This was not easy either – an authentic story demanded a lot of details that could only be obtained from specialist literature.[8]

His choice was right: children would have found adventures in theoretical physics beyond their comprehension. It would have been much easier for him to write on theoretical physics, but this was a subject more suited for adult readers better adjusted to abstract thinking and the fact that physics was an experimental science.

Today popular science is written mostly for grown-ups; there is no doubt that Bronstein could have told a lot about the meaning and drama of scientific research both to those who knew nothing about science and to those who dedicated their lives to it.

It is a much more difficult and responsible task to address the young readers. Bronstein's books belong in Literature with a capital L. Daniil Danin, who writes a lot about popular science, uses *The Solar matter* as a yardstick in this literature [176].

## 6.4. Personality

What sort of a person was this man, endowed with a profound mind and many gifts and who accumulated so much knowledge?

It is not easy to get an idea of what sort of person Bronstein was from eye-witness evidence, out of different "projections": sometimes the impressions are contradictory. This is quite natural, since his was a complicated and harmonious personality more like an object of quantum physics where "projection" depends on the experimental situation as a whole. Let us try to recreate him by using pairs of opposite qualities.

Anselm remembered Bronstein as both self-assured and modest [96]. Having learned that Anselm was going on a boat trip along the Dnieper, Bronstein asked to be taken along. Anselm answered jokingly: "Can you swim? I would hate being responsible to science should you drown". Bronstein said: "Don't you worry, I am no Landau; I'm more of a teacher than a scientist".

He believed that Landau's potential was greater than his own. At the same time he was sure enough of himself and his knowledge to say: "contrary to what such authorities as Niels Bohr and Paul Dirac believe, I am of an opposite opinion" [81, p. 218].

A. Migdal who was Bronstein's post-graduate student in the last year of his life recalls that his teacher seemed to him strong and weak at the same time. In everyday life, say in an overcrowded tram, he felt an urge to support and protect the frail Bronstein from the pushing people. During scientific discussions, however, when he swiftly moved to the blackboard or commented in a witty and logical way, people tended to forget about his slight built and stammering. In his essay [134], V. Berestetsky called him "The Little One". People less interested in science and prone to judge people by their appearance would have thought it comic to see Bronstein boldly attacking his much larger opponents; it would have amazed them that Bronstein himself seemed unaware of his small height.

He could not boast good health either – he came down with pneumonia several times. He did his best, however, to match his "*mens sana*" with a "*corpore sano*": he tried tennis, rowing, swimming and cycling.

He was both respectable and childish. Many of his close friends addressed him by his name and patronymic (usually reserved in Russia for less intimate friends and business acquaintances). Nobody found it strange – he became mature at a very early age and was never embarrassed by his "adult" manners. He was polite to the point of being old-fashioned; he never remained seated in the presence of a standing woman, be she his wife's friend or domestic help. From childhood he learned always to be neatly dressed (a tie and a three-piece suit were his usual outfit), carefully combed and shaven. (All this helped the nickname of Abbot stick.)

Yet there was another Bronstein who could be relaxed and playful. There is a photo showing him holding a teddy-bear, or with a kerchief on his head. As a secretary of the 1933 Nuclear Conference he had a small frog drawn on his official badge (it might be famous Ehrenfest's proverbial frog, which had jumped in the most important moments of discussion).

D. Ivanenko still keeps a postcard written in Bronstein's hand and posted on November 6, 1934, from Samarkand, where he was lecturing together with Krutkov. It said:

> Inshallah! Salam! Dymus, leaving Samarkand for Bukhara we saw this postcard at the station and remembered you. [The postcard reproduced Vatagin's picture of a gorilla.] Not grudging expense we bought and mailed it. We remain your friends as ever,
> Y. Krutkov, M. Bronstein. The 9th of Ramadan, 1354 anno Hegirae

Bronstein was too devoted to science and the quest for truth to tolerate unfounded claims of profound knowledge. He could abandon his polished manners and even be caustic and malicious. He sometimes would skillfully prove some proposition and then, having obtained agreement from a hapless seeker of a scientific glory, immediately disprove it, to the delight of his audience. In the puppet show that he wrote in the wake of the 1933 Nuclear Conference, he ridiculed all the speakers.

He never believed in pleasing everybody; his out-of-the-ordinary personality stirred different feelings in different people, as is always the case. Some were irritated by his wide and profound knowledge, others felt out of place when he demonstrated his free-thinking and unbounded irony. In fact, some of his victims nursed grudges for quite a long time. Yet he was critical and ironic about himself as well, so many of his close and not-so-close friends cherish the image of a tactful and tolerant man. These properties would leave him in the face of bellicose ignorance and dogmatism – he could be equally slashing at a scientific conference, in a tram or at the office of a publisher. Naturally enough, those on the receiving end thought him ill-willed.

His students, on the other hand, retained the image of an indulgent teacher pleasantly surprised at the minutest trace of knowledge and lavish with high marks. This was not indifference; rather it was an awareness that you would not implant physics by force and that encouragement could carry the teacher much farther.

At the same time, Bronstein felt responsible for the future of physics in his country and the fate of the young people he had to teach. He was always unequivocal with those who had no special gifts as theoreticians and were just walking along a well-trodden path. Indeed, it was hard to recognize this in a conscientious young man endowed with a fair share of natural abilities. He was the direct opposite of Y. Frenkel who, in his unbounded kindness, seemed to believe that anybody could become a theoretician [139, p. 121]. There was kindness in Bronstein's approach, too: the earlier the student realized that he was unsuited for the chosen trade the better his chances were to find his true vocation.

Everybody bears the stamp of his occupation. Theoreticians cannot do without theorizing on everyday matters. Landau was one such case. Bronstein also showed an inclination to be over-rational about what people did and felt. This made his judgments sharp (those less acquainted with him believed him to be cynical). In actual fact he was a cynic in the sense Antisthenes and Diogenes had been preaching, that is, the cynic's refusal to comply with generally accepted norms rested on a profound contemplation about the meaning of the "generally accepted".

He was aware of the fact that too much logic was damaging and deadly; therefore he was careful not to extend his rationalism too far. He could be attentive and tender – on one occasion he wrote to a friend with an expectant wife: "I wish her easy lambing" and sent a basket of flowers to the young mother. He could be caustic with friends and yet praised with the words: "This is a true friend who will never let you down".

An intensive intellectual life and theoretical abstractions form a sort of wall between a man and "real life". Theory for Bronstein was life, not just office hours, yet with him boldness of thought and word were never detached from the real world.

People are united in admiring Bronstein's moral purity that was never despotic – in fact, he was always tolerant. (Landau believed that Bronstein carried his tolerance too far.) This was not indiscriminate kindness – it was, rather, a deep-rooted conviction that "it takes all kinds to make the world". He tended to be lenient with those who overstepped moral precepts being unaware of them, like a child would do, and he was harsh with those who did it deliberately.

He never acted against his convictions; he was a friend to two physicists who were rabid enemies. They could barely cope with this unnatural situation yet neither of them was able to persuade Bronstein to drop the other.

He valued the truth above all else. Here is one such case.

There are eye-witness accounts of how Bronstein defended his doctorate in 1935. Vladimir Fock, his former university lecturer, acted as his official opponent. He highly assessed the thesis and offered an opinion on the situation in theoretical physics in general. Bronstein, who thought differently, vehemently attacked him and even dropped the usual polite expressions. He did not agree with Fock on the question that he had been pondering for a long time and felt it unnecessary to stifle his disagreement. After a while, it was hard to tell who was defending.

One should bear in mind that the two men held each other in high esteem; they were colleagues as well as friends. Both worked at the Physicotechnical Institute and lectured at University where Fock headed the department of quantum mechanics while Bronstein was acting head of the theoretical physics department.

He was a citizen in the most lofty meaning of the word, witness his letter to Fock dated April 1937:

> I am of the opinion (even if it is of doubtful origin) that "social welfare is higher than the welfare of private individuals". It beats me when I hear that N. should be allowed to teach mechanics to

> physicists only because he is a nice man and is in need of money .... He regards teaching as a duty a scientist is supposed to pay to the state to be able to support himself and do research in his free time. This approach is dishonest and wrong irrespective of whether practiced by a good or bad scientist. [99]

Fock was the first to come to Bronstein's home after his arrest to find out what had happened. In March 1939 he sent a letter to A. Vyshinsky, the USSR Attorney-General, together with Bronstein's reference signed by S. Vavilov, L. Mandelstam and I. Tamm, and Marshak's letter. Here is what Fock wrote:

> I support the request of Lydia Korneevna Chukovskaya to revise the case of her husband, former assistant professor at Leningrad University, Matvei Petrovich Bronstein.
> He has revealed himself as a talented young scientist who made a great contribution to Soviet science and who is exceptionally well-versed in theoretical physics. His doctorate on Einstein's General Relativity theory produced valuable scientific results. He has obtained new and valuable results on the theory of metals and semiconductors. Finally, he has written several popular science books for young people, all of which were favorably reviewed in central press.
> In case you find it possible to meet L.K. Chukovskaya's request will you please take into account that Bronstein is a valuable scientist.

We are not able to go into details of what happened to Matvei Bronstein in August 1937. By that time arrests had ceased to be something out of the ordinary: in the autumn of 1936 N. Kozyrev and Yu. Krutkov were arrested. Both were too close friends to allow the thought that the accusations heaped on them were true. On August 1, 1937, people with a search and arrest warrant turned up at Bronstein's flat. They destroyed all his manuscripts, looked through his books and confiscated them. He was away, having gone for his annual leave.

He was arrested in Kiev, at his parents', where he had stayed for some time. It all happened in the small hours of the morning. He laughed when asked to surrender his arms and poison. He left home with a towel and asked his mother not to return his train ticket since he expected to be back soon ...

They drove him to Leningrad. Somebody saw him being led along the platform with the towel around his neck. In February 1938, having waited for the *n*th time for several hours, his wife learned the verdict – *ten years of hard labor in a far-away concentration camp without the right to write and receive letters and with all his belongings confiscated.* At that time nobody knew that this was a euphemism for *death by firing squad.*

It was in December 1939 that they managed to learn that he was dead.

Twenty years later, when Matvey Bronstein was rehabilitated, his widow learned the date of his death – February 18, 1938.

# Afterword

It is painful to sum up a life cut short in its prime. Did Matvei Bronstein manage to do much during his thirty years on Earth? The list of his works, the number of physicists whom he taught and a veritable multitude of readers speak for themselves. And yet much was left undone – he had just approached the most fruitful age. One can only wonder how the quantum theory of gravitation would have developed if he was allowed to live his natural life. It is distressing to think about the books that were left unwritten.

His last articles hint that he was interested in the fields of quantum theory, cosmology, astrophysics and nuclear physics. One cannot be sure that a stunning discovery was in store for him – besides knowledge, this requires a good deal of luck. One feels absolutely definite, however, that he would have made a great contribution into physics because an ability to analyze and catalyze physical ideas, just as a talent for teaching, are less affected by outward factors.

Those who knew Bronstein believe that, had he lived, his personality, his scientific authority, his readiness to speak out when necessary and genuine integrity would have improved the climate in the academic community – his very presence would have curbed base instincts or dictatorial claims and encouraged those who needed it.

Bronstein was mostly concerned with quantum mechanics and the theory of relativity, two pillars of the 20th-century physical picture of the world. In fact, he was the first to see them as two sides of *one and the same* pillar. The indeterminacy principle, the linchpin of quantum theory, has already been fitted into the methodology of biographical writing in the Foreword; time has come to do the same with the principle of equivalence, the cornerstone of General Relativity.

It forms the basis of the ideas of geometrization and non-linear interaction, two fundamental principles of physics today. When applied to science's advances within its own space-time, the "geometry" of science is changed by concentrating knowledge and spiritual energy. In fact, scientific evolution and revolutions can be guided only by a non-linear theory: all personal impacts can be taken into account or discarded only to the extent of personal creative energies, so far nobody has suggested a unit for measuring them. In the world of physical phenomena unification is a challenging and attractive aim. This seems to be the opposite of the world of people, where the highest achievements are scored by unique personalities.

Bronstein was one of those born to embellish mankind and illuminate some of the enigmas of nature. He would probably have responded ironically to these words – he never perceived himself as an embellishment or a beacon. The light has reached us after fifty years.

# Afterword to the English Edition

## Half a Century Later

New facts about Bronstein's posthumous life came to light after the Russian edition had appeared.

*1. From the KGB-NKVD Archives*

In summer 1990 Lydia Chukovskaya was allowed to look into her husband's inquisition file. Obviously she was aware that it was no more than a pile of paper sheets sticked together by NKVD/KGB officers – they had nothing to do with the true story of her husband's last months. Yet there was no other evidence. Not even a grave ...

The file opens with an arrest warrant issued on 1 August, 1937 and an order by the Kiev State Security department of 5 August that said:

> M.P. Bronstein who is trying to escape arrest should be detained for an active involvement in Leningrad counterrevolutionary organization.

He was arrested at night in Kiev in his parents' flat. In prison they took from him a voucher to a Kislovodsk sanatorium, a soap dish, a toothbrush, shoelaces ...
The People's Commissar of Ukraine for Internal Affairs ordered:

> Bronstein Matvei Petrovich arrested as a dangerous criminal should be despatched to Leningrad to the NKVD Leningrad Regional Department in an individual compartment of a prison van.

Out of the documents in the file only one was indubitably done in Bronstein's own hand – it was a questionnaire of 15 August. There is one genuine signature of his that confirmed the minutes of the interrogation of 2 October when he rejected all accusations heaped on him. Other signatures cannot be recognized as done in his own hand.

At that time the interrogators relied more on their imagination than on what they could extract from the people they had to deal with. Their zeal and ambitions rather than reality fed their imagination. This can be clearly seen from another inquisition file – Lev Landau's file. (He was arrested in April 1938.) The minutes of interrogation that Landau signed is a careless mixture of facts and stupid inventions. Here is one of the examples of the official's zeal: he asked Landau whether he had informed Bronstein about a leaflet they planned to distribute in April 1938. According to the document Landau answered that he had failed to tell Bronstein about it [1]. In actual

L.Landau in NKVD prison, 1938

fact Landau was only too well aware that his colleague had been arrested long before April 1938.

The minutes of Bronstein's interrogations are nothing more than his interrogator's "flights of fantasy". According to one of them dated 9 October Bronstein was a member of a

> counterrevolutionary organization of intelligentsia who wanted to topple down Soviet power and set up a new political order that would allow intelligentsia to take part in state administration together with the other social groups according to the Western pattern" and was working "to create a basically fascist state that would be able to resist communism". Besides, according to the same author, Bronstein "was resolutely opposing materialist dialectics being applied to natural science".

On 2 December, at the next interrogation session, he was confronted with supporting

> individual terror against the leaders of the CPSU(B) and the government as the only efficient form of struggle.

On 16 December inquisition was completed.

According to the indictment signed on 24 January 1938 the NKVD

> exposed and liquidated a fascist terrorist organization that had been set up in 1930–32 by the German intelligence ...In 1933 it contacted the Trotsky-Zinoviev organization in Leningrad". Bronstein's "practical anti-Soviet activity" consisted in "preparing terrorist acts against the CPSU(B) leadership and the Soviet Government", he also "did a lot of harm to geological prospecting and melioration"; he was a "foreign spy" and supplied "theoretical substantiation of terror as the only correct form of anti-Soviet struggle.

On 18 February 1938 the Military Judicial Board of the Supreme Court of the USSR sat from 8:40 till 9:00 a.m. The verdict was: "Death by a firing squad and confiscation of all personal property". It should have been immediately executed. There is a note in the file stating that the verdict was carried out.

The same file contains pleads of his colleagues physicists and writers who tried to help him never suspecting that he was dead. Earlier like letters seemed to go into other files or into garbage-can.

There are also documents of the period of Bronstein's rehabilitation including those about the investigators Georgi Karpov, Nikolai Lupandin and their chief Yakov Shapiro.

Shapiro met his death from a firing squad in 1939 during the "anti-Ezhov purge". In August 1938, Lupandin, a sadistic torturer, was exiled to the NKVD department of economic management that was believed to suit more his lack of schooling. Poet Nikolai Zabolotsky who had a misfortune to meet him in his office shortly after Bronstein's execution described him at great length [2]. In 1977 this worthy man was allowed to retire on a privileged pension [3]. Karpov made even a more spectacular career – he climbed up to the post of the Chairman of the Council of the Russian Orthodox Church Affairs at the USSR Council of Ministers. His "feats" of 1937 were punished with a reprimand twenty years later [4].

The last page in the 90-page file was dated 1958:

> L.K. Chukovskaya should be rewarded for the binocular taken away from her flat during the search on 1 August 1937.

## 2. *The Last Days in the Cell*

Late in 1990, after her *Notes about Anna Akhmatova* appeared Lydia Chukovskaya got a call from Boris Velikin, her contemporary. He had just read Chukovskaya's diary in which she described how she had first met the great Russian poet. Their roads crossed at the Leningrad prison where Akhmatova regularly came to enquire about her son and Chukovskaya about her husband.

> ...It was in February 1938. I leaned over to look into the wooden embrasure in the Shpalernaya Street and said: 'Bronstein. Matvei Petrovich'. A deep bass answered from somewhere above: 'Taken away'. The man whose face I could not see pushed back with his elbow and fat stomach my hand clutching the money. [5]

At that point Velikin realized who was the man he had met in a prison cell in December 1937 that remained engraved in his memory for the rest of his life.

Velikin himself was arrested on 4 December and brought to the prison in Shpalernaya. A workaholic from the Kirov Plant and a man dedicated heart and soul to Soviet power he was shocked to find himself in a cell designed for 16 and packed with more than a hundred. Few lucky ones slept in canvass beds suspended from the ceiling; the rest slept on the floor, the newcomers had to be satisfied with a place near the lavatory pan.

Out of hundred with whom he shared the cell Velikin remembered only three or four: an actor of the Moscow Art Theatre who was to play Stalin ten years later; director of the Scientists' Club who could talk about cinema for hours; a railwayman who remained alive thanks to a misprint. Matvei Petrovich stood aside in his memory.

After the prison in Shpalernaya Velikin was sent to a concentration camp on the Kolyma where he worked in a mine. He had to pull mine carts underground. All the time he was aware of life draining out of him. Several times he thought he was dying – he survived by a sheer miracle. He worked on a construction site in Magadan and spent years in Norilsk where he stayed until 1956 [6].

I met him when he was 85, unbent by age and misfortunes: he had two books on metallurgy to his name and was still active as the Chief Expert of a ministry: he had just returned from a inspection trip to the Urals.

Why did he cherish the memory of a man who had shared a cell with him and a hundred of others for two months? Why did it never fade away after many years of terrible experience? He was struck by a remarkable concentration of intellect, ramified culture and high morality.

Few of them felt like talking about crazy accusations, they tried to escape into the human world of work, poetry and cinema through lectures and quizzes. Matvei Bronstein earned applause with his lecture on the relativity theory – yet a mysterious subject at that time. It was to be expected: after all he was a physicist. Besides, he proved able to answer any question in any field well beyond the scope of theoretical physics and knew more poetry by heart than anybody else in the cell. What struck Boris Velikin more was Bronstein's ability to penetrate deep into the essence of phenomena. This ability allowed Bronstein to explain to him, who was a professional metallurgist, the subtleties of the special steel technology. There was another man who shared Velikin's admiration – before the revolution he had improved the gun design but it was Bronstein who explained to him the genuine nature of his invention.

This was Bronstein's calling and profession – to explain the essence of things.

Was he able to look deep into the social nightmare he was caught in? He never discussed his case; it seemed that he had no premonition of what was in store for him.

Was his the heaviest burden? Alexander Witt from Moscow and Semen Shubin from Sverdlovsk, two talented young physicists, were arrested at the same time [7]. They were sentenced to five and eight years of forced labor respectively – both died in the winter of 1938 on the Kolyma. As if this fate was not cruel enough they had to live through being transported there together with criminals and through many other hair-raising experience described by those who were lucky to survive.

*3. Subnuclear Physics, Matvei Bronstein and Ettore Majorana*

This happened in July 1991 in the ancient Sicilian city of Erice during the 29th International School of Subnuclear Physics conducted by the Ettore Majorana Center for Scientific Culture. It discussed "Physics at the Highest Energy and Luminosity: to Understand the Origin of Mass".

Even before it was opened the spirit of history permeating the proceedings. The school was sponsored by the World Federation of Scientists and the World Laboratory jointly with the Galileo Galilei Foundation and was dedicated to the 400th anniversary of the first great discovery of modern science: 1591–1991. Author: Galileo Galilei. Its formula was quoted on the school's poster $m_i = m_g$. Translated into the language of contemporary physics it means the principle of equivalence.

The school's program suggested that the history of physics was nothing more than a mere decoration designed to amuse the subnuclear physicists.

Indeed, what's quantum chromodynamics, 200 TeV and the string vacua, the puzzling abbreviations LEP and FNAL to Galilei or what's he to them?

In fact, only the titles of the first and the last lectures sounded understandable for Galilei's contemporaries: "The Problem of the Mass: from Galilei to Higgs" read by Lev Okun from Russia and "The Origins of the Mass" read by Higgs himself.

It was Okun who stirred an interest in Bronstein.

The very title of his lecture suggested that he should employ a strong optical device to see the subject, that is, the *cGh*-object-lens described in detail in the book above. This means to investigate the structure of theoretical physics with the help of three fundamental constants: *c*, the velocity of light, *G*, the gravitational constant and *h*, the Planck constant. This allows to introduce the three-dimensional *cGh*-system of coordinates into the "physical theories space" and to contemplate its past, present and even future.

In May 1991 in Moscow at the first Sakharov Conference Lev Okun spoke about the results of such contemplation and mentioned Matvei Bronstein as the man who had introduced this idea back in the thirties [8]. Few of the participants knew the name of this theoretician and children's writer – his life proved to be too short.

Bronstein was also the first to probe into quantizing gravity and to combine the elementary particles physics with cosmology. In fact, these were the steps towards the so far unconquered *cGh* summit and subnuclear physics of the future.

Lev Okun had no intention to mention Bronstein's name (little known in the West) in Erice yet the audience fascinated with the simple and profound *cGh* approach wanted to know more about its author. They heard a sad story about his life, work and tragic death and about his widow Lydia Chukovskaya who exhibited

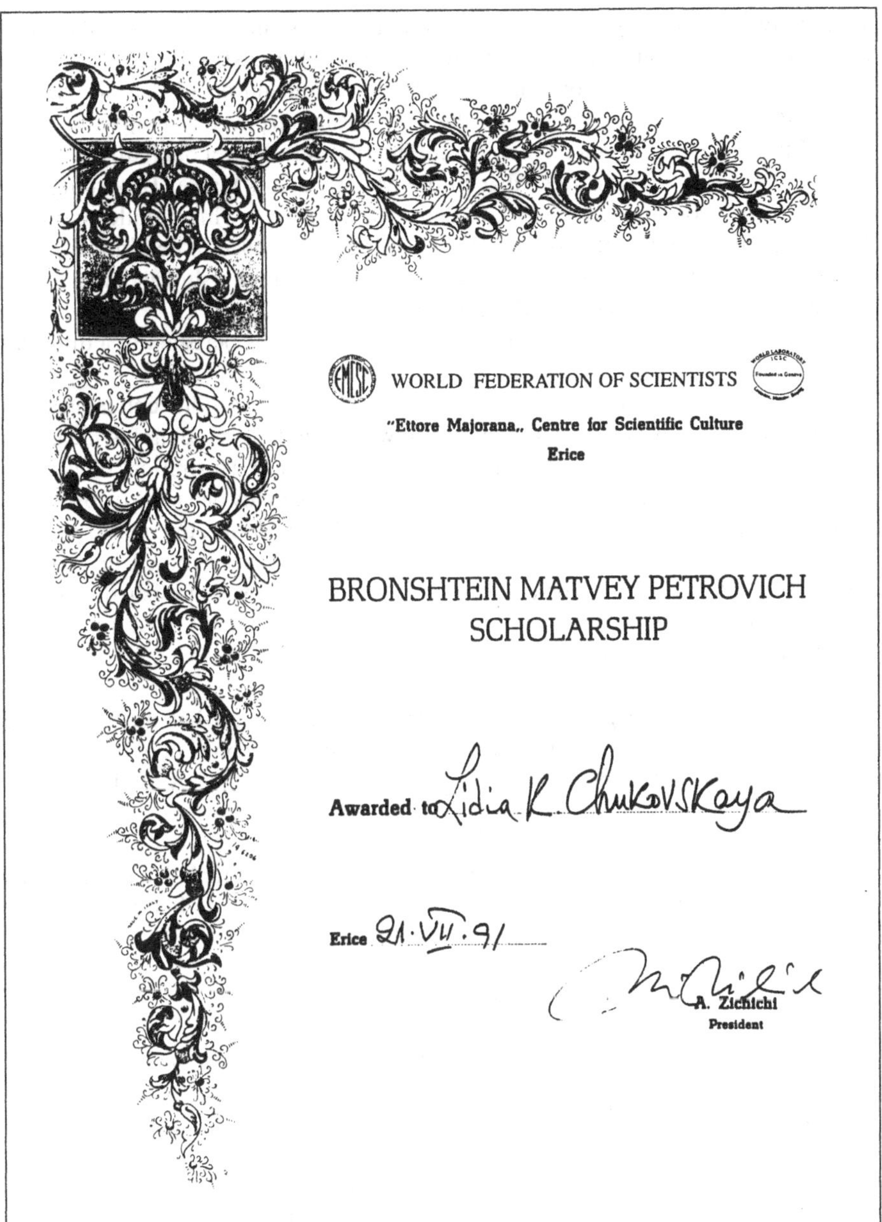

WORLD FEDERATION OF SCIENTISTS
"Ettore Majorana" Centre for Scientific Culture
Erice
BRONSHTEIN MATVEY PETROVICH
SCHOLARSHIP
Awarded to Lidia K. Chukovskaya
Erice 21.VII.91
A. Zichichi
President


The certificate of the first M.P.Bronstein scholarship awarded to Lydia Chukovskaya.

a lot of civic courage and staunchness, about her books and articles first published in the West and her attempts to preserve the memory her husband.

The impression was enormous – the school command decided to set up the Matvei Bronstein scholarship. He was the second Russian physicist to give his name to the Ettore Majorana center scholarship. The first was Andrei Sakharov.

It is impossible to pass over that Ettore Majorana and Matvei Bronstein both were born in 1906 and died in 1938, both had no graves. This parallel is rather superficial: being enigmatically indifferent to life and steeped in pessimism gifted Italian choose to take his own life. In March 1938 he boarded a ship and no one saw him again. In what abyss did he disappear? It is unknown.

Matvei Bronstein brimming with life and creative plans perished in a different abyss.

Majorana's mysterious disappearance dealt a heavy blow to his colleagues and friends who did their best to heal the gaping wound with developing his ideas and writing about him [9]. They gave his name to the Center of Theoretical Physics.

In his country Bronstein's name remained a taboo for many decades. It cost Lydia Chukovskaya a lot of efforts to reissue in 1959 his *Solar Matter*, the masterpiece of literature about science for teenagers, – fear was still deeply rooted in people's minds. In 1965 his second book for children *The X Rays* was published again; in 1990 his last book, *The Inventors of Radiotelegraph,* appeared – back in 1937 its first edition was destroyed.

Lydia Chukovskaya got the first scholarship named after her husband – the colorful document adorns a wall in her flat side by side with a portrait of her husband and photographs of other people who have become part of her life: Andrei Sakharov, Boris Pasternak, Alexander Solzhenitsyn, Anna Akhmatova ...

Let us hope that in Bronstein's homeland they will also establish a prize named after him that can mark achievements in two widely removed fields: in the theory of quantum gravity and popular science for children.

*Additional Bibliography for the English Edition*


1. Gorelik G.E., "My Anti-Soviet Activity ..." (A Year in Landau's Life), *Priroda,* 1991, No. 11, pp. 93–104.
2. Zabolotsky N., A Story of My Imprisonment, *Daugava,* 1988, No. 3, pp. 107–115.
3. Lunin E., The Great Soul, *The Leningrad Panorama*, 1988, No. 5, pp. 24–38.
4. *Rehabilitation. Political Processes of the 30s–50s,* Moscow, Politizdat, 1991, p. 80.
5. Chukovskaya L., *Notes about Anna Akhmatova,* Book 1, 1938–1941, Moscow, 1989.
6. Velikin B.A., Interviews of January 29, 1991 and March 7, 1991.
7. Gorelik G.E., They Had No Time to Become Academicians, *Priroda,* 1990, No. 1, 123–128; *Repressed Science,* Leningrad, Nauka Publishers, 1990, pp. 333–349.

8. Okun L.B., The Fundamental Physical Constants, *Uspekhi fizicheskikh nauk*, 1991, Vol. 161, No. 9, pp. 177–194.
9. Amaldi E., *La vita e l'opera di Ettore Majorana*, Rome. 1966.
10. Bronstein M.P., *The Solar Matter. The X Rays. Inventors of Radiotelegraph*, Moscow, Nauka, 1990 (The Kvant Library series, Issue 80.)

# Chronology

Dec. 2, 1906 – born in Vinnitsa (Ukraine)
1925 – first scientific papers (on the photon structure of X-rays)
1926 – entered Leningrad University
1929 – works on astrophysics, first popular-science book
1930 – graduated from the university and started working in the theoretical department of the Leningrad Physico-Technological Institute
1932 – works on the theory of semiconductors
1934 – first artistic-scientific book *The Solar Matter*
Nov., 1935 – defense of the doctorate "Quantisation of Gravitational Waves"
1936 – papers on quantisation of the weak gravitational field
1937 – paper on the possibility of the photon's spontaneous decay that substantiated the fact of the expanding Universe
Aug. 6, 1937 – arrest at the parental home in Kiev
Feb. 18, 1938 – death sentence by the Military Tribunal of the USSR Supreme Court and death

# Notes

## Notes to Chapter 1

1 Petr Tartakovsky was born in Kiev in 1895 into the family of a prominent doctor, S. Tartakovsky, who was a professor at Kiev University and headed a hospital. Upon graduation from Kiev University in 1918 Petr became a post-graduate student at his alma mater. His first papers dealt with quantum theory. In 1925 he moved to Leningrad on Abram Ioffe's invitation. He headed a laboratory at the Physicotechnical Institute and lectured in the Polytechnical Institute. In 1929 he went to Tomsk to set up the Siberian Physicotechnical Institute. He made his major contribution to science with experiments on crystal diffraction of slow electrons (1927) and on the photoeffect in dielectrics (the thirties). His surveys and popular books and articles were important in disseminating quantum ideas in the country. He died in 1940 of a heart attack.

2 The book was based on a lecture course Tartakovsky gave in autumn 1926 at the Department of Physics and Mechanics of the Leningrad Polytechnic.

3 L. Kordysh's (1878–1932) reference was of special importance. A university professor and corresponding member of the Ukrainian Academy of Sciences, he took a lively interest in the latest developments in physics. Before the 1917 revolution he visited Planck's and Poincaré's laboratories. He was the first in Russia to write about General Relativity: his paper appeared in 1918 [206]. Paradoxically enough, he was able to acquaint himself with the General Relativity theory at an early stage thanks to the German occupation of Kiev, when German scientific journals were regularly sent over to the Ukraine. Physicists outside Kiev had to wait some years to get access to foreign publications on physics.
Kordysh gave much of his time and strength to teaching. He was the first to set up an evening secondary school for young workers; in 1923 he was teaching at the specialized secondary school at which the Bronstein brothers were students for a short while. He could have picked Matvei at that time and suggest the university physics circle to him.

4 A. Goldman (1884–1971) set up and headed the Kiev Research Department of Physics, later transformed into the Kiev Research Institute of Physics. He was a professor at the Kiev Polytechnic and an academician of the Ukrainian AS from 1929 on.

## Notes to Chapter 2

1 Those who graduated from the workers' department had the right to enter the university without exams. Later, about three-fourths of the enrollment dropped out.

2 A. Anselm, E. Kanegisser, V. Kravtsov and I. Sokolskaya were also members of the Jazz-band.

3 S. Kostinsky (1867–1963), corresponding member of the USSR AS, the senior astronomer with the Pulkovo Observatory, was also a university lecturer on photographic astronomy. D. Martynov described him as a "smug pedant" inclined to "pompous banalities". The issue he was hunting for offered some quite original interpretations of recent events in astronomy and characteristics of those involved.

4 E. Hopf (b. 1902), an outstanding German mathematician, was working at that time at Berlin University. The Wiener-Hopf equation and method were a result of his studies on the mathematical side of the radiation shift.

5 There is another evidence of cooperation between Ambartsumyan and Bronstein found in the

additions to the proofs (dated December 1930). It is a more exact assessment obtained by Ambartsumyan. In a special long note to the article, Milne deemed it necessary to note that the value for $q(0)$ was obtained by Bronstein and Hopf independently, that he had received the journal with Bronstein's article and Hopf's manuscripts in one batch of post. (Hopf's article later appeared in the *Monthly Notices*) and that, therefore, priority belonged to Bronstein.

6 Solomon Reiser (1905–1990) was a literary scholar, specialist in text studies, and professor at the Leningrad Institute of Culture.

## Notes to Chapter 3

1 In June 1985, R. Peierls visited the Soviet Union; he showed some of the photographs taken fifty-five years before at the Odessa Congress; in one of them Pauli, Frenkel, Tamm and Simon continued their discussion dressed in bathing trunks on a beach; in others young physicists were enjoying themselves on the sandy beach as all young people would do in similar circumstances.

2 This popular science journal was intended for well-informed readers; its editorial board comprised Ioffe, Kagan, Shpolsky and Schmidt, who was the editor-in-chief. In 1931, a larger journal *Sorena* (a Russian abbreviation for "Socialist Reconstruction and Science") was opened in its stead.

3 Bronstein wrote that "the idea of discrete space emerged many centuries ago and had very little in common with physics. In fact, in the Middle Ages, Jewish and Arabian theologians were disputing whether space was discrete or continuous (see Kurt Lasswitz's *Geschichte der Atomistik*)". He then continued that "the forerunners of the new discrete geometry were little interested in such trifles as relativistic invariance".

4 Dirac's book *The Principles of Quantum Mechanics*, which was translated by Bronstein, appeared in 1932. Ivanenko edited it. Dirac brought English proof sheets to Kharkov. It requires no wisdom to guess that Frenkel advised Bronstein to spare his trousers by not sitting for too long at a desk. Bronstein admired the subtle mind and charm of Frenkel's wife. The general tone of the letter is eloquent evidence of the very specific relationship between the two generations.

5 This paraphrases two lines popular in Germany at the time about General Relativity theory. *Die Quantelung* is written instead of *das Kruemungsmass*.

6 The prehistory also includes the idea of the "atoms of time" voiced by Poincaré at the first stage of the quantum theory [168, p. 27].

7 One of the main occupations of Talmudists was to formulate laws based on the "first principles" of the Bible to suit the ever-emerging new phenomena.

8 The difference between the physical and mathematical ideas of the world is best illustrated by the case of Weyl's unified theory. He was obviously shocked by the unenthusiastic response from physicists who failed to appreciate his approach to the idea of locality that was more consistent than Einstein's "half-hearted" approach realized in General Relativity [170]. History demonstrates again and again that mathematical consistency may lead to a physical inferno.

9 Frederiks and Fock also considered Friedmann to be a mathematician. Bronstein reminded of Einstein's "blunder", which caused him to think that Friedmann was wrong. The Lemaitre model of 1927 did not contain principally new elements when compared to Friedmann's model, still the pioneer of non-static cosmology sank into oblivion. Einstein was one of the few who paid tribute to him.

10 Though the head of a team that studied liquid crystals Frederiks has a place of his own in the history of General Relativity in this country. When the first World War started he found himself in Germany working under Hilbert in Göttingen. Interned as a Russian subject he continued to work under Hilbert as his assistant. Late in 1915 he witnessed the birth of General Relativity with Hilbert's active participation. As soon as the political situation stabilized, Frederiks returned to Russia,

where he became the most active promoter of relativist ideas. In 1921 he published the first Russian review on GR. Together with Friedmann they began a definitive course *Foundations of Relativity*, the first and only part of which appeared in 1924 [131].

11 At that time experiments on rejuvenation were very popular. Voronov was one of the experimenters. Landau wrote to Kapitza on November 25, 1931:
"Dear Petr Leonidovich, it is absolutely necessary to elect Jonny Gamow into the Academy. There is no doubt that he is the best theoretician in the USSR. I am very much afraid that Abrau (Ioffe) is jealous enough to oppose him in one way or another. Really, somebody should restrain the old man. Will you be so kind as to write a letter to the Academy's secretary praising Jonny. Please, send me a copy so that I could publish it in *Pravda* or *Izvestia* together with letters from Bohr and others. This would be splendid if you could ask the Crocodile to do the same! Yours, Landau".
Judging by his reply, the Crocodile [Rutherford] was left alone. (We thank P.Rubinin for the materials from Kapitza's archive.)

12 The work written jointly by Bronstein and Frenkel in 1930 was ample evidence of this [9]. One can surmise that they discussed the Rabi result [256] – emergence of electron levels in a magnetic field. Frenkel explained it in the simplest way possible by using Bohr's postulates. In fact, his style of thinking demanded nothing more – the simplest model and the simplest calculation method. Bronstein found this wanting – he believed that quantizing (if it existed at all and did not depend on relativism in the way Rabi presented it) should be deducted from Schrödinger's equation. The solution of this problem not only affirmed the "simplest" result but also specified and substantiated it. These two approaches are a good illustration of the differences in their styles.

13 In 1933 there were about 50 nuclear physicists in the Soviet Union. (Today one laboratory may employ as many.) There were about 100 personal invitations sent to those who might be interested in nuclear physics. In actual fact, there were even more people who wanted to come.

14 Soon after the conference there was a puppet show at the Physicotechnical Institute based on the recent developments in physics. The play was written by Bronstein; after the show the puppets were presented to the prototypes [159].

15 The book bears traces of this battle: the table of contents shows a section, "From the Translators", absent from the book and does not show the sections "From the Publishers" and "From the Editor" that can be found in the book. Ivanenko was repressed in 1935; by the time the letter was written he was employed by the Physicotechnical Institute in Tomsk.

# Notes to Chapter 4

1 The authors are grateful to Rudolf Peierls for his comments to his article [247].

2 The reference is to the so-called nitrogen catastrophe. The nucleus's properties depend on the parity of the number of its particles. In pre-neutron days a nuclei composed of $k$ neutrons and $l$ protons were considered to contain $k + l$ protons and $k$ electrons. Generally speaking, the parity of the $k + l$ and $2k + l$ numbers do not coincide. This is true, in particular, of nitrogen.

3 This is the place to remind the reader that Dirac formulated his hypothesis on the gravitational constant dependent on cosmological time in 1937 [170]. One can even surmise a connection between these ideas. Dirac attended the first All-Union Nuclear Conference in 1933 at which Bronstein spoke on cosmological problems (see Section 3.10).

4 Probably because Landau had been arrested in April 1938, Pauli's Russian article ascribed this breakthrough to Einstein, rather than to Landau.

5 This could have been connected, in some way, with the Russian edition of Jacobi's *Lectures on Dynamics* that appeared late in 1936. The general "symmetry-conservation" interconnection (established by Nöther in 1918) had not yet been accepted as part of the physical arsenal.

6 Landau referred to him as "an ignorant scribbler" [215]; Ioffe wrote about the pertness with which Lvov penned his articles for one of the most popular thick journals [196]. In the thirties, *Novy mir* regularly carried Lvov's surveys under the general heading "On the Physical Front".

7 Lvov was inventive in his heated defense of the conservation principle: for example, he never mentioned Bohr and ascribed the non-conservation hypothesis to "Kramers, Slater and company".

8 The second article by the same authors [111], which appeared in late 1934, eight months after Fermi's theory had been generally recognized, was of a different nature. They kept strictly within physics – scientific arguments needed no ideological crutches.
Some thirty-five years later the same Blokhintsev was prepared to recognize that the law of energy conservation was violated "in the world of elementary particles, especially in the sphere of high energies"; he further noted that "it is hard to tell between these violations and the processes in which the neutral particles are involved" [107, p. 309].
One can easily imagine how indignant young Blokhintsev would have been provided he glimpsed some of what he would be writing at 62. He provided an answer in his *Contemplations on the Problems of Cognition* that appeared after his death: "Society should be reasonably sure of its ideals to allow new thoughts and ideas that would outstrip the old confines. It should be patient to take time for their assessment" [108, p. 56]. He also admitted that "the laws of conservation of energy and momentum could not be applied to a young universe".

9 The authors are grateful to S. Vonsovsky who shared with them his reminiscences about Shubin. Shubin's background was partially responsible for his ideological stand. His father, a legal expert and journalist, was on the *Pravda* editorial board and worked in the Komintern; his brother-in-law was one of the founders of the Komsomol and the Communist Youth International.

# Notes to Chapter 5

1 Recently there appeared some weighty reasons to believe that due to some concealed causes gravitation has an important role to play in the structure of the microworld. Half a century ago this was a matter of faith, Einstein was its prophet.

2 Weyl also presupposed gravitation's statistical nature in the hopes of explaining the weakness of gravity with a huge number of particles in the universe [170].

3 The fact that the time of the gravitational radiating away has a more than cosmological value reminds us that at the same time Einstein was pondering the cosmological problem. He did not attach much significance to the effect's value that was connected with the idea of a static universe he supported at that time. In the static (i.e., eternal) Universe the effect of atomic instability is inadmissible regardless of its value. This was quite natural for its time only – in our time the proton's possible instability is regarded as preferable [262]. This is how the evolutionary cosmology affected the norms of the admissible in theoretical physics.

4 His gratitude to Pauli "for numerous critical comments and suggestions" points to the connection with [159].

5 The reference to Planck relates the text both to the past and the future. In 1899 Planck introduced (for purely metrological purposes) natural units based on the constants $c$, $G$ and the just recently proposed $\hbar$; it first came to light in Bronstein's thesis of 1935 that the same Planck units coincided with quantum limits of GR (see Section 5.4).

6 The very name *Journal of Russian Physico-Chemical Society* had an old-fashion and provincial flavor. Indeed, this journal was born in 19th century when Russian physics was very provincial. New wine needed new bottle, and in 1931 *ZhRFKhO* was replaced be *ZhETF, Journal of Experimental and Theoretical Physics*, where $\hbar$ and $c$ were much more common.

7 In his article [21], while referring to a work by Landau and Peierls, Bronstein thanked Landau and

emphasized that "quite a few ideas that underlie the present article were prompted by their discussions".

8 In 1926 R. Fowler attributed white dwarfs' greater density to their being composed of degenerate fermi-gas. It was Frenkel who was the first in 1928 to discuss relativistic fermi-gas as applied to the theory of super-dense stars [290].

9 Today, one would find it hard to comprehend Landau's attitude to the less "pathological" sources of stellar energy, including the thermonuclear synthesis of hydrogen in helium: "It would be strange if high temperatures could assist in the process only because they promote some chemical processes (chain reactions!)". To understand the physics of that time one should accept this skepticism (explained in Chapter 4).

10 When describing his joint work with Dirac and Podolsky in the newspaper *Tekhnika* (March 18, 1936) Fock wrote: "The ideas that underlie this theory were successfully applied to gravitation by Leningrad physicist M. Bronstein; he derived the Newtonian law of gravitation out of the ideas of the 'gravitational quanta'".

11 It was in 1937 that Gamow first mentioned the possibility of this connection in an attempt to squeeze dry the idea of paired exchange forces. Since the interaction caused by the $e\nu$ pair exchange turned out to be $10^{12}$ times weaker than it should be to make the nucleus explainable, Gamow surmised that the nuclear forces were caused by the $e^+e^-$ pair exchange, that is, that exchange $ee$-interaction was $10^{12}$ times stronger than the $e\nu$-interaction. From this it followed that the exchange $\nu\nu$-interaction was still $10^{12}$ times weaker than the $e\nu$-interaction. This explained why the emission of neutrino pairs was associated with the quantum emission of gravitational radiation [150]. As recently as 1962 Gamow wrote that "a pair of neutrinos could have produced spin 2" (that is, the graviton) [162]. One can see that the gravitational explanation was a sort of a by-product, while electromagnetism and gravitation's geometrical nature were left outside. Such a way of inventiveness was alien to Bronstein.

12 There is no negation in the original; this is an obvious misprint.

13 Naturally enough, Bronstein was concerned with the theory's justifications; not only with its inner perfection. As he defended the thesis Frederiks asked him how gravitational radiation could have betrayed itself. Bronstein pointed at the effect of slowing rotation of binary stars [173, p. 319]. It is worth comparing this answer, given in 1935, with the first observational proof of such radiation discovered in 1979 in a binary pulsar.

14 In 1917, when Einstein was thinking about his first cosmological model, he regarded the uniformity of matter distribution as a hypothesis while the relatively small velocities of stars (!) as "the most important thing of all we know about the distribution of matter" [306].

15 Dirac was an exception that cannot be ignored. In 1937 he suggested a speculative project that subordinated physics' fundamental principles to cosmology. He presented, in particular, the gravitational constant $G$ as a function of time, $G(t)$. (About reasons for this see [170]. He found little sympathy among his colleagues. In any case there was no place for a variable gravitational "constant" $G(t)$ in Bronstein's $cG\hbar$ map.

# Notes to Chapter 6

1 It was Bohr, author of the principle of complementarity, who said: "*Contraria non contradictoria sed complementa sunt*".

2 He studied Spanish by reading *Don Quixote* in a tram while riding for about an hour from his home to the institute.

3 Lydia Chukovskaya saved this letter, which had been sent to the high offices with a request to revise the verdict. We feel that the kind and flattering words about Bronstein were not prompted

by the situation. Twenty years later Chukovsky wrote in his diary: "Painfully familiar situation in Russia – a stifled talent. [Then he gave the names of Russian poets and writers, among them S.Esenin, O.Mandelstam, I.Babel, M.Tsvetaeva and Mitya Bronstein] – all of them were trampled down by the same boot".

4 When asked by a teenager what people should strive for, Bronstein smiled: " I don't know. One should strive for what one wants more than anything else. Yet I know better what you should avoid: stagnation of thought".

5 E. Kanegisser wrote to R. Peierls: "Dau is absolutely downcast .... Don't know what to do about this .... Now the Abbot and he are the closest friends – it seems that they will never quarrel" [224, p. 43].

6 The second edition of *Statistical Physics* was typeset in February 1939, its foreword being dated May 1939.

7 There were administrators at Detgiz Publishers who tried to prevent a new edition of *The Solar Matter* despite these flattering words.

8 Bronstein, who read practically everything that appeared on physics, probably found his task less taxing. There is a well-known book on the history of Roentgen's discovery written by Glasser. It first appeared in Germany and later, in 1933, its extended edition appeared in Britain [165]. Bronstein was the first to borrow it from his institute's lending library. The second was Röntgen's pupil A. Ioffe, the institute's director.

# Bibliography

*Abbreviations*

PZM – Pod znamenem marxisma [Under the Banner of Marxism]
PZS – Physikalische Zeitschrift der Sowjetunion [published in Kharkov between 1932 and 1938].
UFN – Uspekhi fizicheskikh nauk [Advances of Physical Sciences]
ZhETF – Zhurnal Eksperimentalnoy i Teoreticheskoy Fiziki [Journal of Experimental and Theoretical Physics]
ZhRFKhO – Zhurnal Russkogo Fiziko-Khimicheskogo Obchshestva [Journal of the Russian Physico-Chemical Society]
ZP – Zeitschrift für Physik

## Works by Matvei Bronstein

### *Scientific Papers and Surveys*


1. Ob odnom sledstvii gipotezy svetovykh kvantov [About one consequence from light quanta hypothesis] // ZRKhRO, 1925, Vol. 57, pp. 321–325.
2. Zur Theorie des kontinuierlischer Röntgenspektrums // ZP, 1925, Bd 32, S. 881–885.
3. Bemerkung zür Quantentheorie des Laue-Effektes // Ibid., S. 886–893.
4. Über die Bewegung eines Elektrons in Felde eines festen Zentrums mit Berücksichtigung der Massenveränderung bei der Ausstrahlung // ZP, 1926, Bd 35, S. 234, 863; Bd 39, p. 901.
5. Zur Theorie der Feinstruktur des Spektrallinien // ZP, 1926, Bd 37, S. 217–224.
6. Zum Strahlungsgleichgewichtsproblem von Milne // ZP, 1929, Bd 58, S. 696–699.
7. Über das Verhaltnis des effektiven Temperatur der Sterne zur Temperatur ihrer Oberfläche // Ibid., Bd 59, S. 144–148.
8. K teorii obchshei tsirkulyatsii atmosfery [On theory of general circulation of atmosphere] // Zhurnal geofiziki i meteorologii, 1929, Vol. 6, pp. 265–292.
9. Kvantovanie svobodnykh elektronov v magnitnom pole [Quantizing of free electrons in magnetic field] (with Y. Frenkel) // ZhRFKhO, 1930, Vol. 62, pp. 485–494.
10. On the temperature distribution in stellar atmospheres // Mon. Not. Roy. Astron. Soc., 1930, Vol. 91, p. 133.
11. Sovremennoe sostoyanie relativistskoi kosmologii [Modern state of relatovistic cosmology] // UFN, 1931, Vol. 11, pp. 124–184.
12. On the theory of electronic semoconductors // PZS, 1932, Vol. 2, pp. 28–45.
13. Fizicheskie svoistva elektronnykh poluprovodnikov [Physical properties of electron semoconductors] // Zhurnal Tekhnicheskoy Fiziki, 1932, pp. 919–952.
14. On the anomalous scattering of gamma-rays // PZS, 1932, Vol. 2, p. 541.
15. Poglochshenie i rasseyanie gamma-luchei [Absorbing and scattering of gamma-rays] // UFN, 1932, Vol. 12, p. 649.
16. On the expanding universe // PZS, 1933, Vol. 3, pp. 73–82.
17. On the conductivity of semiconductors in magnetic field // Ibid., p. 140.

18. Vnutrennya konversiya gamma-luchei [Inner conversion of gamma-rays] // UFN, 1933, Vol. 13, p. 537.
19. Vsesoyuznaya yadernaya konferentsia [All-union nuclear conference] // Ibid., p. 768.
20. Vnutrennee stroenie zvezd i istochniki zvezdnoi energii [Inner composition of stars and soerces of stellar enrgy] // Uspekhi astron. nauk. Coll. 2, Moscow, ONTI, 1933, pp. 84–103 (see also [50, pp. 142–166]).
21. K voprosu o vozmozhnoi teorii mira kak tselogo [On the question of the possible theory of the world as a whole] // Ibid, Coll. 3, Moscow, ONTI, pp. 3–30; [50, pp. 186–215].
22. The second law of thermodynamics and the Universe (with L. Landau) // PZS, 1933, Vol. 4, pp. 114–118.
23. On the applicability limits of the Klein-Nishina's formula // PZS, 1924, Vol. 5, p. 517.
24. K voprosy o relativistskom obobchshenii printsipa neopredelennosti [On the question about relativistic generalisation of indeterminancy principle] // DAN. 1934, Vol. 1, pp. 388–390.
25. Svoistva izlucheniya pri ochen vysokikh plotnostyakh energii [Properties of radiation under very high energy densities] // Ibid., Vol. 2, p. 462.
26. O konferentsii po teoreticheskoi fizike [About the conference on theoretical physics] // UFN, 1934, Vol. 14, pp. 516–520.
27. O rasseyanii neitronov protonami [On the scattering of neutrons by protons] // DAN, 1935, Vol. 8, p. 75.
28. Gipotezy o proiskhozhdeii kosmicheskikh luchei [Hypotheses about the origin of cosmic rays] // Trudy Vses. konf. po izycheniyu stratosfery, Leningrad; Moscow, 1935, pp. 429–432, 445–449.
29. Addition to the traslation: Einstein A. Osnovy teorii otnositel'nosti, Moscow-Leningrad, ONTI, 1935.
30. Quantentheorie schwacher Gravitationsfelder // PZS, 1936, Vol. 9, pp. 140–157.
31. Kvantovanie gravitatsionnykh voln [Quantizing of gravitational waves] // ZhETF, 1936, Vol. 6, pp. 195–236
32. On the anomalous scattering of electrons by protons // PZS, 1936, Vol. 9, p. 537.
33. On the intensity of forbidden transitions // Ibid., p. 542.
34. Ueber den Spontanen Zerffall der Photonen // PZS, 1936, Vol. 10, pp. 686–688.
35. O vozmozhnosti spontannogo rachsheplenia fotonov [On possibility of spontaneous splitting of photons] // ZhETF, 1937, Vol. 7, pp. 335–358.
36. O magnitnom rasseyanii neitronov [On the magnetic scattering of neutrons] // Ibid., pp. 357–362.

## *Encyclopaedical Entries*

37. Otnositel'nosti teoriya [Theory of relativity] (with V. Frederix) // Tekhnich. entsiklopedia, Vol. 15, Moscow, Gostekhteorizdat, 1931, pp. 352–367.
38. Elektron // Ibid., Vol. 26, 1934, pp. 645–650.
39. Atom // Ibid., Additional vol., 1936, pp. 78–97.
40. Atom // Fizich. slovar', Vol. 1, Moscow, ONTI, 1936, pp. 214–222.
41. Beta-luchei spektry. Beta-raspada teoriya [Beta-rays theory] // Ibid., pp. 298–302, 307–313.
42. Kvantovaya statistika [Quantum statistics] // Fizich. slovar', Vol. 2, 1937, pp. 744–751.

## *Reviews*

43. *Dirac P.*, The principles of quantum mechanics. (Oxford, 1930) // UFN, 1931, Vol. 11, pp. 355–358.
44. *Weyl H.*, Gruppentheorie und Quantenmechanik (2 Aufl. Lepzig, 1932) // Ibid., pp. 358–360.

45. *Gamow G.A.*, Stroenie atomnogo yadra i radioaktivnost' [The constitution of atomic nuclei and radioactivity] // UFN, 1932, Vol 12, p. 362.
46. *Joos G.*, Lehrbuch der theoritischen Physik (Leipzig, 1932), PZS, 1933, Vol. 3, pp. 100–101.
47. *Terenin A.N.*, Vvedenie v spektroskopiyu [Introduction in spectroscopy] (Leningrad, 1933) // UFN, 1934, Vol. 14, p. 248.
48. *Heisenberg W., Shrödinger E., Dirac P.*, Sovremennaya kvantovaya mekhanika. Tri nobelevskikh doklada [Three Nobel lectures] (Leningrad, Moscow, 1934) // PZS, 1934, Vol. 6, pp. 612–615.

## *Editing*

49. *Dirac P.*, Osnovy kvantovoi mekhaniki, Moscow, Gostekhteorizdat, 1932; 1937.
50. Osnovnye problemy kosmicheskoi fiziki, Kharkov-Kiev, ONTI, 1934.
51. *Brillouin L.*, Atom Bohra, Moscow, ONTI, 1934.
52. *Becker R.*, Elektronnaya teoriya, Moscow, ONTI, 1936.
53. *Born M.*, Tainstvennoe chislo 137 // UFN, 1936, Vol. 16, pp. 687–729.

## *Popular Science Books and Articles*

54. Vsemirnoe tyagotenie i elektrichestvo (Novaya teoriya Einsteina) // Chelovek i priroda, 1929, No. 8, pp. 20–25.
55. Sostav i stroenie zemnogog shara. (Popular science library of the Nauka i Tekhnika journal, Issue 77), Leningrad, Krasnaya gazeta, 1929.
56. Yaponski schetny pribor Soroban // Chelovek i priroda, 1929, No. 15, pp. 5–7.
57. Efir i ego rol' v staroi i novoi fizike // Ibid., No. 16, pp. 3–9.
58. Electron i tselye chisla (novye raboty A.S. Eddingtona) // Chelovek i priroda, 1930, No. 2, pp. 8–16.
59. Proikhozhdenie Solnechnoi sistemy // Ibid., No. 23, p. 3–10.
60. O prirode polozhitel'nogo elektrichestva // Nauch. slovo, 1930, No. 5, pp. 91–99.
61. Henri Russel // Tvortsy nauki o zvezdakh, Leningrad, Krasnaya gazeta, 1930, pp. 39–49.
62. James Jeans // Ibid., pp. 75–88.
63. Stroenie atoma // Library of self-education for workers, Book 1, Leningrad, Krasnaya gazeta, 1930.
63a. Budova atoma, Kharkiv-Odessa, 1931 (in Ukrainian).
64. Novy krizis teorii kvant // Nauch. slovo, 1931, No. 1, pp. 38–55.
65. Element s atomnym nomerom 0 // Sorena, 1932, No. 7, pp. 165–167.
66. O prirode kosmicheskikh luchei // Ibid., pp. 142–144.
67. Uchenie o khimicheskoi valentnosti v sovremennoi fizike // Priroda, 1932, No. 10, pp. 875–878.
68. Convegno di Fisica Nucleare // Sorena, 1933, No. 1, pp. 176–177.
69. K voprosu o neitronakh // Priroda, 1933, No. 1, pp. 63–66.
70. Konferentsia po tverdym nemetallichskim telam // Ibid., pp. 73–74.
71. Elektronnye poluprovodniki // Priroda, 1933, No. 2, pp. 54–56.
72. O knige Rutherforda, Chadwika, Ellisa // Ibid., p. 77.
73. Polozhitelnye elektrony // Priroda, 1933, No. 5/6, pp. 21–22.
74. Anomalnoe poglochshenie i rasseyanie gamma-luchei // Ibid., pp. 110–111.
75. Vnutrennyaya konversiya gamma-luchei // Priroda, 1933, No. 8/9. pp. 87–89.
76. Problemy fiziki zvezd // Sorena, 1933, No. 7, pp. 12–23.
77. Vsesoyuznaya yadernaya konferentsiya // Sorena, 1933, No. 9, pp. 155–165.
78. Iskusstvennaya radioaktivnost' // Sorena, 1934, No. 5, pp. 3–9.

79. Sokhranyaetsya li energiya? // Sorena, 1935, No. 1, pp. 7–10.
80. Uspekhi nauki i tekhniki v 1934 g.: Fizika atomnogo yadra // Sorena, 1935, No. 2, pp. 78–81.
81. Stroenie vechshsestva, Leningrad-Moscow, ONTI, 1935 (a fragment of it is found in the Kvant journal 1978, No. 3, pp. 11–18).
82. Atomy, elektrony, yadra, Leningrad-Moscow, ONTI, 1935 (second edition: Atomy i elektrony, Moscow, Nauka, 1980 // Biblioteka Kvant, Issue 1).
82a. Ataka atomnogo yadra, Kiev, 1936 (in Ukrainian).
83. Samyi sil'ny kholod // Ezh, 1935, No. 8, pp. 18–20.
84. Novosti fiziki // Izvestia, May 12, 1936.

### *Artistic-Scientific Books*

85. Solnechnoe vechshestvo // Koster, Coll. 2, Leningrad, Detizdat, 1934; God XVIII. Almanach 8, Moscow, 1935, pp. 413–460/Introduced by S.Y. Marshak; Leningrad, Detizdat, 1936; Moscow, Detgiz, 1959/Introduced by L.D. Landau, afterword by A.I. Shal'nikov.
85a. Sonyachna rechovina, Kharkiv-Odessa, Ditvidav, 1937 (in Ukrainian).
85b. Der Sonnenstoff, Kiev, Ukrderschnazmenwydaw, 1937.
86. Luchi X // Koster, 1936, No. 1, Leningrad, Detizdat, 1937; Moscow, Malysh, 1965.
87. Izobretateli radiotelegrapfa // Koster, 1936, Nos. 4, 5; Kvant, 1987, No. 2 (the first chapters).
87a. Solnechnoe vechshestvo. Luchi X. Izobretateli radiotelegrapfa. Moscow, 1990.

## Other Works

88. *Adamov G.B.*, Kniga o solnechnom vechshestve // Mol. gvardiya, 1937, No. 1, pp. 217–223.
89. Akademik Lev Davidovich Landau, Moscow, Znanie, 1978, 62 pp.
90. Albert Einstein i teoriya gravitatsii, Moscow, Mir, 1979.
91. *Ambartsumyan V.A.*, Vnutrennee stroenie i evolyutsiya zvezd // Mirovedenie, 1934, No. 4, pp. 245–256.
92. *Ambartsumyan V.A.*, Statistika Fermi i teoriya belykh karlikov // *Rosseland S.*, Astrofizika na osnove teorii atoma, Moscow, ONTI, 1936, pp. 140–144.
93. *Ambartsumyan V.A.*, Letter to G.Gorelik of 14.1.1984.
94. *Ambarzumian V.*, *Iwanenko D.*, Zur Frage nach Vermeidung der unendlichen Selbstrückwirkung des Elektrons // ZP, 1930, Bd 64, S. 563–567.
95. *Andronikashvili E.L.*, Vospominaniya o zhidkom gelii, Tbilisi, 1980, pp. 72–73.
96. *Anselm A.I.*, Letter to G.Gorelik of 26.4.1984.
97. *Arzumanyan A.*, Aragats, Moscow, Sov. pisatel', 1979, 312 pp.
98. Archives RAN (Leningrad branch), Record group 970 (Bursian V.R.), Inventory 1, File 83.
99. Ibid., Record group 1034 (Fock V.A.), Inventory 3, File 980.
100. Archives LGU, Record group 7240, Inventory 10, File 74.
101. Ibid., Record group 1, Inventory 3, Box 11, File 387.
102. Archives LPI im. M.I. Kalinina, Personal file 564.
103. Archives LFTI, Personal file 287, p. 12.
104. Ibid., Record group 3, Inventory 2, File 2.
105. Atomnoe yadro (Coll. of papers at the 1st All-Union Nuclear Conf.), Moscow, Gostekhteorizdat, 1934, 216 pp.
106. *Bedny D.*, Do atomov dobralis' // Poln. sobr. soch. Vol. 14, Leningrad, 1930, p. 52.
107. *Blokhintsev D.I.*, Prostranstvo i vremya v mikromire, Moscow, Nauka, 1970, 250 pp.

108. *Blokhintsev D.I.*, Razmyshleniya o problemakh poznaniya i tvorchestva i zakonomernostyakh protsessov razvitiya // Teoriya poznaniya i soveremnnaya fizika, Moscow, Nauka, 1984, pp. 53–74.
109. Dmitri Ivanovich Blokhintsev, Dubna, OIYaI, 1977, 64 pp.
110. *Blokhintsev D.I., Gal'perin F.M.*, Bor'ba vokrug zakona sokhraneniya i prevrachsheniya energii v sovremennoi fizike // PZM, 1934, No. 2, pp. 97–106.
111. *Blokhintsev D.I., Gal'perin F.M.*, Gipoteza neitrino i zakon sokhraneniya energii // Ibid., No. 6, pp. 147–157.
112. *Blokhintsev D.I., Gal'perin F.M.*, Atomistika v sovremennoi fizike // PZM, 1936, No. 5, pp. 102–124.
113. *Bohr N.*, Ueber die Anwendung der Quantentheorie auf den Atombau.I. // ZP, 1923, Bd 13, S.117–165.
114. *Bohr N.*, Ueber die Wirkung von Atomen bei Stossen. // ZP, 1925, Bd 34, S.142–157.
115. *Bohr N.*, Atomic theory and mechanics // Nature, Suppl., 1925, V.116, p.845–852.
116. *Bohr N.*, Atomic stability and conservation laws // Atti del Convegno di fisica nucleare della Fondatione A. Volta, 1931, Rome, 1932, pp. 119–130.
117. *Bohr N.*, Chemistry and quantum theory of atomic constitution // J. Chem. Soc., 1934, p.349–384.
118. *Bohr N.*, Sur la methode de correspondance dans la theorie de l'elrctron // Sept. Conceil de Physique. Inst. Intern. de Physique Solvay. Paris, 1934, p.216–228.
119. *Bohr N.*, Conservation laws in quantum theory // Nature 1936 V.138, p.25.
120. *Bohr N., Kramers H., Slater J.*, The quantum theory of radiation // Phil. Mag., 1924, V.47, pp.785–800.
121. *Bohr N., Rosenfeld L.*, Zur Frage der Messbarkeit der elektromagnetischen Feldgroessen // Kgl. Danske Vidensk. Selskab., Math.-Fys. Medd., 1933, Bd 12, N 8, S.3–65.
122. *Vavilov S.I.*, Referaty knig: Einstein A., Efir i printsip otnosiltel'nosti (Petrograd, 1921); Lenard P., Über Relativitätsprinzip, Ather, Gravitation (1920) // UFN. 1921, Vol. 2, Issue 2, p. 300.
123. *Vavilov S.I.*, Eksperimental'nye osnovaniya teorii otnositel'nosti, Moscow-Leningrad, 1928, 128 pp.
124. *Veinberg S.*, Raspad protona // UFN, 1982, Vol. 137, pp. 150–172.
125. *Veselov M.G.*, Letter to G.Gorelik of 25.5.1984.
126. *Vizgin V.P.*, Razvitie vzaimosvyazi printsipov invariantnosti s zakonami sokhraneniya v klassicheskoi fizike, Moscow, Nauka, 1972, 240 pp.
127. *Vizgin V.P.*, Relatistskaya teoriya tyagoteniya (istoki i formirovanie. 1900–1915), Moscow, Nauka, 1981, 350 pp.
128. *Vizgin V.P.*, Edinye teorii polya v pervoi treti XX veka, Moscow, Nauka, 1985, 300 pp.
129. *Vizgin V.P.*, O newtonovskikh epigrafakh v knige S.I. Vavilova o teorii otnositel'nosti // Newton i filosofskie problemy fiziki XX v., Moscow, Nauka, 1989.
130. *Vizgin V.P., Gorelik G.E.*, The reception of the Theory of Relativity in Russia. In: The Comparative Reception of Relativity. Ed. Th.Glick. Boston, 1987. P.256–326.
131. *Vizgin V.P., Frenkel V.Y.*, Vsevolod Kosntantinovich Frederix – pioner relativizma i fiziki zhidkikh kristallov v SSSR // Einsteinovski sbornik, 1984–1985, Moscow, Nauka, 1988, pp. 105–138.
132. Vklad akademika A.F. Ioffe v stanovlenie yadernoi fiziki v SSSR, Leningrad, Nauka, 1980, 32 pp.
133. *Vladimirov Y.S.*, Kvantovaya teoriya gravitatsii // Einsteinovski sbornik, 1972, Moscow, Nauka, 1974, pp. 280–340.
134. *Volkov Vl., (Berestetski V.B.)*. Seminar // Puti v neznaemoe, Col. 15, Moscow, Sov. pisatel', 1980, pp. 423–441.
135. *Vonsovski S.V.*, Magnetizm, Moscow, Nauka, 1971, 1032 pp.

136. *Vonsovski S.V.*, Vospominaniya o Semene Petroviche Shubine // Iz istorii estestvoznaniya i tekhniki Pribaltiki (Riga), 1984, Vol. 7, pp. 189–195.
137. *Vonsovski S.V., Leontovich M.A., Tamm I.E.*, Semen Petrovich Shubin (k 50-letiyu so dnya rozhdeniya i 20-letiyu so dnya smerti) // UFN, 1958, Vol. 65, pp. 733–737.
138. Vospominaniya o I.E. Tamme, Moscow, Nauka, 1986, 310 pp.
139. Vospominaniya o Y.I. Frenkele, Moscow, Nauka, 1976, 278 pp.
140. *Vyaltsev A.N.*, Diskretnoe prostrantsvo-vremya, Moscow, Nauka, 1965, 320 pp.
141. *Heitler W.*, The quantum theory of radiation, Oxford, 1936, 290 pp.
142. *Halpern O.*, Scattering processes produced by electrons in negative energy state // Phys. Rev., 1933, Vol. 44, pp. 855–856.
143. *Gamow G.A.*, Ocherk razvitiya ucheniya o stroenii atomnogo yadra. Teoriya radioaktivnogo raspada // UFN, 1930, Vol. 30, pp. 531–544.
144. *Gamow G.A.*, O preobrazovanii elementov v zvezdakh // Usp. astron. nauk, Coll. 2, Moscow-Leningrad, Gostekhteorizdat, 1933, pp. 72–83.
145. *Gamow G.A.*, Teoriya Diraca i polozhitel'nye elektrony // Sorena, 1933, No. 8, pp. 25–30.
146. *Gamow G.A.*, Mezhdunarodny kongress po stroeniyu atomnogo yadra // Sorena, 1934, No. 1, pp. 16–21.
147. *Gamow G.*, Über den heutigen Stand (20. Mai 1934) der Theorie des beta-Zerfalls // Phys. Ztschr. 1934, Bd 35, S. 533–542.
148. *Gamow G.A.*, Ocherk razvitiya ucheniya o stroenii atomnogo yadra. V. Problema beta-raspada // UFN, 1934, Vol. 14, pp. 389–406.
149. *Gamow G.* Mr. Tompkins in Wonderland, or Stories of *c*, *G* and *h*, Cambridge, 1939, 62 pp. 2nd ed., 1965.
150. *Gamow G.*, Probability of nuclear meson-absorption // Phys. Rev., 1947, Vol. 71, pp. 550–551.
151. *Gamow G.*, The creation of the Universe, London, 1961, 210 pp.
152. *Gamow G.*, Gravity. Classical and Modern Views, New York, 1962, 243 pp.
153. *Gamow G.*, Thirty years that shook physics, New York, 1966, 320 pp.
154. *Gamow G.*, My world line. An informal autobiography, New York, 1970, 180 pp.
155. *Gamow G., Teller E.*, Some generalisations of the beta-transformation theory // Phys. Rev., 1937, Vol. 51, 289 pp.
156. *Gamow G., Ivanenko D., Landau L.*, Mirovye postoyannye i predel'nyi perekhod // ZhRFKhO, 1928, Vol. 60, pp. 13–17.
157. *Heisenberg W.*, Die Selbstenergie des Elektrons // ZP, 1930, Bd 65, S. 4–13.
158. *Heisenberg W.*, Physical principles of quantum theory. 1930.
159. *Heisenberg W., Pauli W.*, Zur Quantendynamik der Wellenfelder // ZP, 1929, Bd 56, S.1–61.
160. *Hessen B.M.*, Osnovnye idei teorii otnositel'nosti, Moscow-Leningrad, Mosk. rabochii, 1928, 70 pp.
161. *Hessen B.M.*, Efir // BSE, 1st ed., Vol. 65, 1931, pp. 16–18.
162. *Hessen B.M.*, Sotsial'no-ekonomicheskie korni mekhaniki Newtona, Moscow-Leningrad, Gostelkhteorizdat, 1933, 82 pp.
163. *Ginzburg V.L.*, O fizike i astrofizike, Moscow, Nauka, 1985, 397 pp.
164. *Ginzburg V.L., Kirzhnits D.A., Lyubushin A.A.*, O roli kvantovykh fluktuatsii gravitatsionnogo polya v obchshei teorii otnositel'nosti i kosmologii // ZhETF, 1971, Vol. 60, pp. 451–459.
165. *Glasser O.*, Wilhelm Conrad Röntgen and the history of X-rays, London, 1933, 260 p.
166. *Goldman A.G.*, Fizika na Ukraine v 10-yu godovchshinu Sovetskoi Ukrainy // Visnik prirodoznatstva, 1927, No. 5/6, pp. 257–272 (in Ukrainian).
167. Gorelik G.E., Pervye shagi kvantovoi gravitatsii i planckovskie velichiny // Einsteinovski sbornik, 1978–1979, Moscow, Nauka, 1982, pp. 334–365.
168. *Gorelik G.E.*, First Steps of Quantum Gravity and the Planck Values // Studies in the History of General Relativity. (Einstein Studies. Vol.3). Boston, 1992, pp.364–379.

169. *Gorelik G.E.*, O gumanitarnykh kornyakh fizicheskogo mirovozzreniya Einsteina // Issledovaniya po istorii fiziki i mekhaniki, 1965, Moscow, Nauka, 1985, pp. 75–84.
170. *Gorelik G.E.*, Istoriya relyativistskoi kosmologii i sovpadenie bolshikh chisel // Einsteinovski sbornik, 1982–1983, Moscow, Nauka, 1986, pp. 302–322.
171. *Gorelik G.E.*, Zakony OTO i zakony sokhraneniya // Znanie – sila, 1988, No. 1, pp. 23–29.
172. *Gorelik G.E.*, Dva portreta // Neva, 1989, No. 8, pp. 167–173.
173. *Gorelik G.E.*, *Frenkel V.Y.*, M.P. Bronstein i ego rol' v stanovlenii kvantovoi teorii gravitatsii // Einsteinovski sbornik, 1980–1981, Moscow, Nauka, 1985, pp. 291–327.
174. *Gorky M.*, O temakh (1933) // Sobr. soch., Vol. 27, Moscow, Khudozh. lit., 1954, p. 108.
175. *Graham L.*, The Socio-political roots of Boris Gessen // Social Stud. Sci., 1985, Vol. 15, pp. 705–722.
176. *Danin D.*, Zhazhda yasnosti (Chto zhe takoe nauchno-khudozhestvennaya literatura?) // Formuly i obrazy. Spor o nauchnoi teme v khudozhestvennoi literature, Moscow, Sov. pisatel', 1961, pp. 3–67.
177. *Delokarov K.H.*, Filosofskie problemy teorii otnositel'nosti, Moscow, Nauka, 1973, 206 pp.
178. *Delokarov K.H.*, Metodologicheskie problemy kvantovoi mekhaniki v sovetskoi filosofskoi nauke, Moscow, Nauka, 1982, pp. 223–232.
179. *Jammer M.*, The conceptual development of quantum mechanics. N.Y., 1967.
180. *Dirac P.*, Does conservation of energy hold in atomic processes? // Nature, 1936, Vol. 137, pp. 298–299.
181. *Dolgov A.D.*, *Zel'dovich Y.B.*, *Sazhin M.V.*, Kosmologiya rannei vselennoi, Moscow, MGU Press, 1987, 199 pp.
182. *Zeldovich Y.B.*, Avtobiograficheskoe posleslovie // Izbr. trudy. Chastitsy, yadra, Vselennaya, Moscow, Nauka, 1985, p. 437.
183. *Zeldovich Y.B.*, *Novikov I.D.*, Stroenie i evolutsiya Vselennoi, Moscow, Nauka, 1975, 560 pp.
184. *Zel'manov A.L.*, Kosmologicheskie teorii. I // Astron. zhurn., 1938, Vol. 15, pp. 456–481.
185. *Zel'manov A.L.*, Kosmologiya // Astronomiya v SSSR za 30 let. Moscow, USSR AN Press, 1948.
186. *Zel'manov A.L.*, Kosmologiya // Razvitie astronomii v SSSR, Moscow, Nauka, 1967.
187. *Ivanenko D.D.*, Dopolnenie // Dirac P. Osnovy kvantovoi mekhaniki, Moscow, Gostekhteorizdat, 1932.
188. *Ivanenko D.D.*, Konferentsiya po atomnomu yadru v Leningrade // Front nauki i tekhniki, 1933, No. 10/11, pp. 139–143.
189. *Ivanenko D.D.*, Model atomnogo yardra i yadernye sily // 50 let sovremennoi yadernoi fizike, Moscow, Energoizdat, 1982, pp. 18–52.
190. *Ivanenko D.D.*, *Sokolov A.A.*, Kvantovaya teoriya gravitatsiii // Vestn. MGU, 1947, No. 8, pp. 103–115.
191. *Ivanter B.K.*, K pervoi godovchshine raboty izdatel'stva "Detskaya literatura" // Pravda, 1936, Dec. 28.
192. *Idlis G.M.*, Osnovnye cherty nablyudaemoi astronomicheskoi Vselennoi kak kharakternye svoistva obitaemoi kosmicheskoi sistemy // Izv. Astrofiz. in-ta AN KazSSR, 1958, No. 7, pp. 39–54.
193. *Ioffe A.F.*, Elektronnye poluprovodniki, Leningrad-Moscow, ONTI, 1933, 120 pp.
194. *Ioffe A.F.*, Sovetskaya fizika i 15-letie fiziko-tekhnicheskikh institutov // Izvestiya, 1933, Oct. 3.
195. *Ioffe A.F.*, Razvitie atomisticheskikh vozzrenii v XX veke // PZM, 1934, No. 4, pp. 52–68.
196. *Ioffe A.F.*, O polozhenii na filosofskom fronte sovetskoi fiziki // PZM, 1937, No. 11/12. pp. 131–143.
197. *Ioffe A.F.*, O fizike i fizikakh, Leningrad, Nauka, 1985, 544 pp.
198. *Kaplan S.A.*, Fizika zvezd, Moscow, Nauka, 1970, p. 110.
199. *Kikoin I.K.*, Rasskazy o fizike i fizikakh, Moscow, Nauka, 1986, 144 pp.
200. *Kirzhnits D.A.*, Problema fundamental'noi dliny // Priroda, 1973, No. 1, pp. 38–45.

201. *Kirzhnits D.A., Linde A.D.*, Fazovye prevrachsheniya v micromire i vo Vselennoi // Priroda, 1979, No. 11, pp. 20–30.
202. *Klein M.*, The first phase of the Bohr-Einstein dialogue // Hist. St. Phys.Sci., v.2. Philadelphia, 1970, pp.1–30.
203. *Klein O.*, Zur fünfdimnesionalen Darstellung der Relativitatstheorie // ZP, 1927, Bd 46, S. 188.
204. *Kobzarev I.Y., Berkov A.V., Zhizhin E.A.*, Teoriya tyagoteniya Einsteina i ee eksperementalnye sledstviya, Moscow, MIFI, 1981, 164 pp.
205. *Kobzarev I.Y.*, Predislovie // Einsteinovsky sbornik. 1982–1983, Moscow, Nauka, 1986, p. 6.
206. *Kordysh L.I.*, Gravitatsiya i inertsiya // Universitetskie izv. Kiev, 1918, Vol. 57, No. 3/4, pp. 1–20.
207. *Kordysh L.I.*, Gravitatsionnaya teoriya difraktsionnykh yavlenii // Ibid., pp. 1–36.
208. *Kordysh L.I.*, Teoriya otnositel'nosti i teoriya kvant // Izv. Kievskogo politehknich. i s/kh in-tov, 1924, Book 1, Issue 1, pp. 10–17.
209. *Kochina P.Y.*, Nikolai Evgrafovich Kochin, Moscow, Nauka, 1979, p. 74.
210. *Krum S.*, Nekotorye cherty sovetskoi fiziki // Sorena, 1936, No. 4, pp. 120–124.
211. *Krum S.*, Fizika na sessii Akademii Nauk // Ibid., No. 5, pp. 105–116; 155–162.
212. *Krum S.*, Iskhod noveishego spora o sokhranenii energii // Ibid., No. 8, pp. 85–87.
213. *Landau L.D.*, Diamagnetismus der Metalle // ZP, 1930, Bd 64, S.629.
214. *Landau L.D.*, On theory of stars // PZS 1932, Bd 1, S.285.
215. *Landau L.D.*, Burzhuazia i sovremennaya fizika // Izvestiya, 1935, Nov. 23.
216. *Landau L.D.*, On sources of stellar energy // Nature 1938, V.141, p.333.
217. *Landau L.D.*, Teoriya kvant ot Maksa Planka do nashikh dnei // Maks Plank (1858–1958), Moscow, 1958, pp. 94–108.
218. *Landau L.D.*, Quantum theory of field // Niels Bohr and the development of physics. Ed. W.Pauli. London, 1955.
219. *Landau L.D., Lifshits E.M.*, Statisticheskaya fizika, Moscow, Gostekhteorizdat, 1938, 272 pp.
220. *Landau L.D., Lifshits E.M.*, Statisticheskaya fizika. 2nd ed., Moscow, Gostechteorizdat, 1940, 280 pp.
221. *Landau L.D., Pierels R.*, Erweiterung des Unbestimmtheitsprinzip fur die relativistischer Quantentheorie // ZP 1931, Bd 69, S.56.
222. *Landau L.D., Pyatigoski L.M.*, Mekhanika, Moscow, Leningrad, Gostekhteorizdat, 1940, 203 pp.
223. *Lemaitre G.*, L'Univers en expansion // Rev. Quest. Sci. 1932, No. 11, p. 391.
224. *Livanova A.M.*, Landau, Moscow, Znanie, 1983, 239 pp.
225. *Lvov V.E.*, Perpetuum mobile – poslednee slovo burzhuaznoi fiziki // Novy mir, 1934, No. 5, pp. 224–242.
226. *Lvov V.E.*, Ataka na zakon sokhraneniya energii // Vesn. znaniya, 1934, No. 5, pp. 265–269.
227. *Lvov V.E.*, Dokument voinstvuyuchshego idealisma // Novy mir, 1935, No. 4, pp. 269–272.
228. *Lvov V.E.*, Nauchnoe obozrenie. O kamuflyazhe, o vechnom dvigatele i shutnike "materialiste" iz zhurnala "Sorena" // Ibid., No. 11, pp. 246–252.
229. Lvov V.E., Na fronte fiziki // Ibid., 1936, No. 5, pp. 139–153.
230. *Lvov V.E.*, Na fronte kosmologii // PZM, 1938, No. 7, pp. 137–167.
231. *Lvov V.E.*, Molodaya vselennaya, Leningrad, Lenizdat, 1969, 220 pp.
232. *Mulkay M.*, Science and the sociology of knowledge. London, 1979.
233. *Markov M.A.*, O prirode materii, Moscow, Nauka, 1976, 192 pp.
234. *Martynov D.Y.*, Pulkovskaya observatoriya v gody 1926 – 1933 // Istoriko-astronomicheskie issledovaniya, Issue 17, Moscow, Nauka, 1984, pp. 425–450.
235. *Marshak S.Y.*, Povest ob odnom otkrytii // God vosemnadtsatyi. Almanakh vosmoi, Moscow, 1935.
236. *Marshak S.Y.*, Dom, uvenchanny globusom // Novy mir, 1968, No. 9, pp. 157–181.
237. *Migdal A.B.*, Poiski istiny, Moscow, Mol. gvardiya, 1983, 210 pp.
238. *Migdal A.B.*, Interviyu zhurnalu "Fizika v shkole" // Fizika v shk., 1986, No. 2, pp. 22–26.

239. *Misner C., Thorne K., Wheeler J.*, Gravitation. San Francisko, 1973.
240. *Nasledov D.N.*, Organizator molodoi shkoly // Industrialny, 1940, Oct. 18.
241. Niels Bohr. Zhizn i tvorchestvo, Moscow, Nauka, 1967, 343 pp.
242. *Okun L.B.*, $\alpha, \beta, \gamma \ldots Z$ (elementarnoe vvedenie v fiziku elementarnykh chastits), Moscow, Nauka, 1985, 110 pp.
243. *Ostwals V.*, Izobretateli i issledovateli, Moscow, 1909, 62 pp.
244. *Ostwald V.*, Velikie lyudi, Moscow, 1910, 372 pp.
245. *Peierls* (Kanegisser) *E.N.* Letter to G.Gorelik of 9.3.1984.
246. *Peierls* (Kanegisser) *E.N.* Letter to G.Gorelik of 10.4.1984.
247. *Peierls R.*, Introduction // Bohr N., Collected Works, Vol. 9 (Nuclear physics, 1929–1952), Copenhagen, 1985, pp. 3–90..
248. *Panteleev L.*, Marshak v Leningrade // Izbrannoe, Leningrad, Khudozh. lit., 1967, p. 491.
249. *Pauli W.*, Einige die Quantenmechanik betreffenden Erkundigunsfragen // ZP 1933, Bd 80, S. 573–586.
250. *Pauli W.*, Raum, Zeit und Kausalitaet in der Modernen Physik (1934) // Scientia 1936 V.59, pp.65–76.
251. *Pauli W.*, Zakony sokhramemiya v teorii otnositelnosti i atomnoi fizike [Conservation laws in the theory of relativity and atomic physics] (1937) // Fizika. Problemy, istoriya, lyudi, Leningrad, Nauka, 1986, pp. 217–232.
252. *Pauli W.*, The papers of last years // Theoretical physics in the twentieth century. Ed. M.Fierz and V.Veisskopf. Cambridge, 1960.
253. *Perel V.Y., Frenkel V.Y.*, Dve raboty Y.I. Frenkelya // Fizika i tekhnika poluprovodnikov, 1984, Vol. 18, pp. 1931–1939.
254. *Planck M.*, Izbrannye trudy, Moscow, Nauka, 1975, 540 pp.
255. Problemy sovremennoi fiziki v rabotakh Leningradskogo fiziko-tekhnicheskogo in-ta, Moscow, Leningrad, Izd-vo AN SSSR, 1936, pp. 74–77.
256. *Rabi I.*, Der Freie Elektron in homogen Magnetfeld // ZP. 1928, Bd 49, S. 507–511.
257. *Raiski S.M.*, Neskolko vospominani // Akademik L.I. Mandelshtam. K 100-letiyu so dnya rozhdeniya, Moscow, Nauka, 1979, p. 215.
258. *Reiser S.A.*, Osnovy tekstologii, Leningrad, Prosvechsenie, 1978, 210 pp.
259. *Rosenfeld L.*, Ueber die Gravitationswirkungen des Lichtes // ZP, 1930, Bd 65, S.589–599.
260. *Rosenfeld L.*, On quantization of fields // Nucl. Phys., 1963, Vol. 40, pp. 353–356.
261. *Savel'ev V.Y.* Letter to V.Frenkel of 5.1.1984.
262. *Salam A.*, Kalibrovochnoe obedinenie fundamental'nykh sil [Nobel lecture] // Na puti k edinoi teorii polya, Moscow, Znanie, 1980.
263. *Semkovski, S.Y.*, Dialecticheski materialism i printsip otnositel'nosti, Moscow, Leningrad, 1926, 216 pp.
264. Sessiya AN SSSR 14–20 marta 1936 // Izv. AN SSSR. OMEN, 1936, No. 1/2.
265. *Sominski M.S.*, Abram Fedorovich Ioffe, Moscow, Nauka, 1964, 643 pp.
266. *Stoner E.C.*, The equilibrium of dense stars // Philos. Mag., 1930, Vol. 9, pp. 944–963.
267. *Stoner E.C.*, A note on condenced stars // Ibid., 1931, Vol. 11, pp. 986–995.
268. *Tamm I.E.*, Teoreticheskaya fizika // Oktyabr i nauchnyi progress, Vol. 1, Moscow, APN, 1967, p. 170.
269. *Tartakovski P.S.*, Ob osnovnykh gipotezakh teorii kvantov // Izv. Kiev. un-ta, 1919, No. 1/2, pp. 1–12.
270. *Tartakovski P.S.*, Kvanty sveta, Leningrad, GIZ, 1928, 240 pp.
271. *Tartakovski P.S.*, Eksperimental'nye osnovaniya volnovoi teorii materii, Moscow, Gostekhteorizdat, 1932, 280 pp.
272. *Tunitski Z.*, Istoriya solnechnogo vechshestva // Koms. pravda, 1936, July 22.
273. *Wheeler J.*, Geons // Phys. Rev., 1955, V.97, p.511–536.

274. *Feinberg E.L.*, Dve kul'tury. Intuitsia i logika v iskusstve i nauke.Moscow, Nauka, 1992.
275. Fock *V.A.*, Uspekhi sovetskoi fiziki // Tekhnika, March 18, 1936.
276. *Fock V.* [summaries of M.P. Bronstein's articles [30, 31]], Zentralblatt für Math. und ihre Grenzgebiete, 1936, Bd 14, S. 87.
277. Fock V.A., Osnovy kvantovoi mekhaniki i granitsy ee prilozhimosti, Moscow, 1936, 38 pp.
278. *Fock V.A.*, Problema mnogikh tel v kvantovoi mekhanike // UFN, 1936, Vol. 16, pp. 943–954.
279. *Fock V.A.*, Albert Einstein (po povodu 60-letiya so dnya ego rozhdeniya // Priroda, 1939, No. 7, pp. 95–97.
280. *Fock V., Jordan P.*, Neue Unbestimmtheitseigenschaften des elektromagnetischen Feldes // ZP, 1930, Bd 66, S. 206–209.
281. *Fock V.A., Podolski B.*, On the quantization of electromagnetic waves and the interaction of charges on Dirac's theory // PZS 1932, V.1, p.801–817.
282. *France A.*, La rotisserie de la Reine Pedauque. Moscow, 1958.
283. *Frederix V.K.*, Obchshi printsip otnositel'nosti Einsteina // UFN., 1921, Vol. 2, pp. 162–188.
284. *Frenkel V.Y.*, Yakov Ilyich Frenkel, Moscow, Nauka, 1966, 472 pp.
285. Frenkel V.Y., Paul Ehrenfest, Moscow, Atomizdat, 1977, 210 pp.
286. *Frenkel V.Y.*, L.A. Artsimovich v LFTI // Vospominaniya ob akademike L.A. Artsimoviche, Moscow, Nauka, 1981, pp. 157–166.
287. *Frenkel V.Y.*, Pervaya vsesoyuznaya yadernaya konferentsiya // Chteniya pamyati A.F. Ioffe, 1983, Leningrad, Nauka, 1985, pp. 74–94.
288. *Frenkel V.Y., Yavelov B.E.*, Einstein – izobretatel, Moscow, Nauka, 1981, 152 pp.
289. *Frenkel Y.I.*, Teoriya otnositel'nosti, Petrograd, Mysl, 1923, 182 pp.
290. *Frenkel Y.I.*, Primenenie teorii Pauli-Fermi k voprosu o silakh stsepleniya (1928) // Sobr. izbr. trudov, Vol. 2, Moscow, Izd-vo AN SSSR, 1958, pp. 109–121 (4. Sverkhplonye zvezdy, pp. 118–121).
291. *Frenkel Y.I.*, O krizise sovremennoi fiziki // Archives of the USSR AN, Record group 1515, Inventory 2, File 104.
292. *Hetherington N.S.*, Philosophical values and observation in Edwin Hubble's choice of a model of the Universe // Hist. Studies Phys. Sci., 1982, Vol. 13, pp. 41–67.
293. *Hawking S.*, A brief history of time. Bantam Books, London, 1988. Ch. 10.
294. *Hopf E.*, Mathematical problems of radiative equilibrium, Cambridge, 1934, 164 pp.
295. *Chandrasekhar S.*, Stellar configurations with degenerate cores // Observatory, 1934, Vol. 57, pp. 373–377.
296. *Chandrasekhar S.*, The highly collapsed configurations of a stellar mass // Mon. Not. Roy. Astron. Soc., 1935, Vol. 95, pp. 207–225.
297. *Chandrasekhar S.*, Perenos luchistoi energii, Moscow, izd-vo inostr. lit., 1953, pp. 85, 96.
298. *Chukovskaya L.K.*, V laboratorii redaktora, Moscow, Iskusstvo, 1963, 290 pp.
299. *Shmushkevich E.M., Davydov B.I.*, Teoriya elektronnykh poluprovodnikov // UFN, 1940, Vol. 24, pp. 21–61.
300. *Shpolski E.V.*, Eksperimentalnaya proverka fotonnoi teorii rasseyaniya // UFN, 1936, Vol. 16, pp. 458–466.
301. *Shubin S.P.*, O sokhranenii energii // Sorena, 1935, No. 1, pp. 11–13.
302. *Shankland R.*, An apparent failure of the photon theory of scattering // Phys. Rev., 1936, Vol. 49, pp. 8–13.
303. *Eigensohn M.S.*, Bol'shaya Vselennaya, Moscow, Gostekhteorizdat, 1936, 142 pp.
304. *Einstein A.*, The investigation of ether state in magnetic field (1895) // Phys.Blaetter 1971, Ausg.9, p.388.
305. *Einstein A.*, Naeherungsweise Integration der Feldgleichungen der Gravitation // Koenig. Preuss. Akad. Wiss. (Berlin). Sitzungsberichte. 1916, pp.688–696.

306. *Einstein A.*, Kosmologische Betrachtungen zur allgemeinen Relativitaetstheorie // Koenig. Preuss. Akad. Wiss. (Berlin). Sitzungsberichte. 1917, pp.142–152.
307. *Einstein A.*, Ueber Gravitationswellen // Koenig. Preuss. Akad. Wiss. (Berlin). Sitzungsberichte. 1918, pp.154–167.
308. *Einstein A.*, Aether und Relativitaetstheorie. Springer Verlag, Berlin, 1920.
309. *Einstein A.*, Ueber den Aether // Schweiz. naturforsch. Gesellschaft, Verhandlungen, 1924, Bd 105, S.85–93.
310. *Einstein A.*, Das Raum-, Feld- und Aetherproblem in der Physik // Die Koralle, 1930, Bd 5, S.486–487.
311. Ehrenfest – Ioffe. Nauchnaya perepiska, Leningrad, Nauka, 1973, 309 pp.

# Appendix 1.
# Extract from M.P.Bronstein's paper

"Quantentheorie schwacher Gravitationsfelder", 1936

# Physikalische Zeitschrift der Sowjetunion

Herausgegeben vom Volkskommissariat für Schwerindustrie der UdSSR.




## Inhalt



Fortsetzung auf der nächsten Umschlagseite.

# QUANTENTHEORIE SCHWACHER GRAVITATIONSFELDER.[1]

*Von M. Bronstein.*

(Eingegangen am 2. Januar 1936.)

§ 1. Allgemeines. § 2. Hamiltonsche Form und ebene Wellen. § 3. Vertauschungsrelationen und Eigenwerte der Energie. § 4. Ein wenig Gedankenexperimentieren! § 5. Wechselwirkung mit der Materie. § 6. Energieübertragung durch die Gravitationswellen. § 7. Herleitung des Newtonschen Gravitationsgesetzes.

## § 1. Allgemeines.

Die Abweichungen eines raumzeitlichen Kontinuums von der „Euklidizität" können bekanntlich durch die Komponenten des Riemann - Christoffelschen Tensors charakterisiert werden. Wenn diese Abweichungen klein sind, kann dieses Tensorfeld vierter Stufe aus einem symmetrischen Tensorfeld zweiter Stufe auf folgende Weise abgeleitet werden:

$$(\mu\rho\nu\sigma) = \frac{1}{2}\left(\frac{\partial^2 h_{\mu\nu}}{\partial x_\rho \partial x_\sigma} + \frac{\partial^2 h_{\rho\sigma}}{\partial x_\mu \partial x_\nu} - \frac{\partial^2 h_{\mu\sigma}}{\partial x_\rho \partial x_\nu} - \frac{\partial^2 h_{\rho\nu}}{\partial x_\mu \partial x_\sigma}\right), \qquad (1)$$

wo $h_{\mu\nu}$ ist die kleine Abweichung des fundamentalen metrischen Tensors von seinem Minkowskischen Wert $\Delta_{\mu\nu}$ ($\Delta_{00} = 1$, $\Delta_{11} = \Delta_{22} = \Delta_{33} = -1$, $\Delta_{\mu\nu} = 0$, wenn $\mu \neq \nu$). Unter diesen Umständen betrachten wir die Welt als eine „Euklidische" mit dem metrischen Tensor $\Delta_{\mu\nu}$ und die $(\mu\rho\nu\sigma)$ als die Komponenten eines in dieser ebenen Welt eingebetteten Tensorfeldes vierter Stufe. Dabei spielen die $h_{\mu\nu}$ die Rolle der „Potentiale", deren Werte durch vier zusätzliche „Eichungsbedingungen"

$$[\alpha\alpha, \beta] = 0 \qquad (\beta = 0, 1, 2, 3) \qquad (2)$$

fixiert werden können. (Hier, und im folgenden, bezieht sich die Summationsvorschrift nur auf griechische Indizes,

[1] Eine ausführlichere Fassung dieser Arbeit erscheint gleichzeitig im „Journal f. exp. und theor. Phys." (russisch).

genügen. Dann sind bekanntlich die Eigenwerte von $\xi\xi^+$ gleich $n+1$, und die von $\xi^+\xi$ gleich $n$, wo $n$ eine positive ganze Zahl oder Null ist.

Hier gelingt es nicht, solche $\xi$-Variablen auf symmetrische Weise einzuführen. Eine mögliche Lösung des Problems sieht so aus

$$\frac{1}{2}(h_{00,\mathfrak{k}} + \sum_l h_{ll,\mathfrak{k}}) = \sqrt{\frac{h}{2\omega d\mathfrak{k}}}\, \xi_{00,\mathfrak{k}}\, e^{i\omega t},$$

$$h_{11,\mathfrak{k}} = \sqrt{\frac{h}{2\omega d\mathfrak{k}}}\left(\frac{\xi^+_{11,\mathfrak{k}}}{\sqrt{3}}\, e^{-i\omega t} + \frac{\xi_{22,\mathfrak{k}}}{\sqrt{3}}\, e^{i\omega t} + \xi_{33,\mathfrak{k}}\, e^{i\omega t}\right),$$

$$h_{22,\mathfrak{k}} = \sqrt{\frac{h}{2\omega d\mathfrak{k}}}\left(\frac{\xi^+_{11,\mathfrak{k}}}{\sqrt{3}}\, e^{-i\omega t} + \frac{\xi_{22,\mathfrak{k}}}{\sqrt{3}}\, e^{i\omega t} - \xi_{33,\mathfrak{k}}\, e^{i\omega t}\right),$$

$$h_{33,\mathfrak{k}} = \sqrt{\frac{h'}{2\omega d\mathfrak{k}}}\left(\frac{\xi^+_{11,\mathfrak{k}}}{\sqrt{3}}\, e^{-i\omega t} - \frac{2}{\sqrt{3}}\, \xi_{22,\mathfrak{k}}\, e^{i\omega t}\right),$$

$$h_{lm,\mathfrak{k}} = \sqrt{\frac{h}{2\omega d\mathfrak{k}}}\, \xi_{lm,\mathfrak{k}}\, e^{i\omega t}, \quad (l \neq m),$$

$$h_{0l,\mathfrak{k}} = \sqrt{\frac{h}{2\omega d\mathfrak{k}}}\, \xi^+_{0l,\mathfrak{k}}\, e^{-i\omega t}.$$

Die Hamiltonsche Funktion wird in den neuen Variablen zu

$$\left.\begin{aligned} H = \sum_k h\omega\,(&\xi^+_{00,\mathfrak{k}}\xi_{00,\mathfrak{k}} + \xi_{12,\mathfrak{k}}\xi^+_{12,\mathfrak{k}} + \xi_{23,\mathfrak{k}}\xi^+_{23,\mathfrak{k}} + \xi_{13,\mathfrak{k}}\xi^+_{13,\mathfrak{k}} - \\ &- \xi^+_{01,\mathfrak{k}}\xi_{01,\mathfrak{k}} - \xi^+_{02,\mathfrak{k}}\xi_{02,\mathfrak{k}} - \xi^+_{03,\mathfrak{k}}\xi_{03,\mathfrak{k}} - \xi_{11,\mathfrak{k}}\xi^+_{11,\mathfrak{k}} + \\ &+ \xi^+_{22,\mathfrak{k}}\xi_{22,\mathfrak{k}} + \xi^+_{33,\mathfrak{k}}\xi_{33,\mathfrak{k}}). \end{aligned}\right\} \quad (5')$$

Darum sind die Eigenwerte der Energie (für jeden Wert von $\mathfrak{k}$)

$$h\omega\,(n_{00} + n_{12} + n_{23} + n_{31} - n_{01} - n_{02} - n_{03} - n_{11} + n_{22} + n_{33} + 2),$$

wo $n_{00}$, $n_{12}$, ... zehn Quantenzahlen sind ($n = 0, 1, 2, \ldots$).

Die Bedingungen (6) machen diesen Ausdruck positiv-definit. Um das einzusehen, betrachten wir wieder den Fall $\mathfrak{k} \parallel z$. Aus (6′) erhalten wir dann folgende Bedingungen:

$$\xi_{00,\mathfrak{k}}\, e^{i\omega t} + \xi^+_{03,\mathfrak{k}}\, e^{-i\omega t} = 0, \qquad \xi^+_{01,\mathfrak{k}}\, e^{-i\omega t} + \xi_{13,\mathfrak{k}}\, e^{i\omega t} = 0,$$

$$\xi^+_{02,\mathfrak{k}}\, e^{-i\omega t} + \xi_{23,\mathfrak{k}}\, e^{i\omega t} = 0, \qquad \xi^+_{11,\mathfrak{k}}\, e^{-i\omega t} + \xi_{22,\mathfrak{k}}\, e^{i\omega t} = 0.$$

Es folgt daraus, dass

$$n_{01} = n_{31} + 1, \quad n_{02} = n_{23} + 1, \quad n_{22} = n_{11} + 1, \quad n_{03} = n_{00} + 1.$$

Die Eigenwerte der Energie für diesen $\mathfrak{k}$ werden zu

$$h\omega\,(\overset{+}{\xi}_{12,\mathfrak{k}}\,\xi_{12,\mathfrak{k}} + \overset{+}{\xi}_{33,\mathfrak{k}}\,\xi_{33,\mathfrak{k}}) = h\omega\,(n_{12} + n_{33}).$$

Wir sehen also, dass die Energie des Gravitationsfeldes aus den positiven Gravitationsquanten besteht, und zwar von je zwei Polarisationen für jeden Wellenvektor $\mathfrak{k}$. Analog zu dem klassischen Fall spielen auch hier nur die transversalen Gravitationsschwingungen eine Rolle: z. B. für $\mathfrak{k} \parallel z$ eine $h_{12}$- und eine $\frac{1}{2}(h_{11} - h_{22})$-Schwingung.

Dabei entstehen keine „Nullpunktsenergieglieder": das ist infolge einer zweckmässig gewählten Reihenfolge der Faktoren im Ausdruck (5) gelungen.

## § 4. Ein wenig Gedankenexperimentieren!

Um den physikalischen Inhalt der Quantentheorie des Gravitationsfeldes etwas näher zu verstehen, betrachten wir die Messung einer der hier vorkommenden Feldgrössen, z. B. des Christoffelschen Dreiindizessymbols $[00,1]$. Die klassischen Bewegungsgleichungen Einsteins lauten in unserem Fall (alle $h_{\mu\nu} \ll 1$):

$$\frac{d^2x}{dt^2} = \frac{\partial h_{01}}{\partial t} - \frac{1}{2}\frac{\partial h_{00}}{\partial x} = [00,1]. \tag{9}$$

Nach dem Vorbilde von Bohr und Rosenfeld[1] betrachten wir die Messung eines raumzeitlichen Mittelwertes von $[00,1]$ im Volumen $V$ und in dem Zeitintervall $T$. Nehmen wir einen Probekörper vom Volumen $V$. Seine Masse sei $\rho V$. Die obere Bewegungsgleichung, die nur dann gilt, wenn die Geschwindigkeit des Probekörpers klein gegenüber der Lichtgeschwindigkeit ist, macht die folgende Messung möglich: messen wir den Impuls des Probekörpers am Anfang und dann am Ende eines Zeitintervalls $T$, so ist der gesuchte

[1] N. Bohr und L. Rosenfeld, Dansk. Vidensk. Selskab., Math.-fys. Meddel. 12, 8, 1933.

Mittelwert definitionweise gleich

$$\frac{(p_x)_{t+T}-(p_x)_t}{\rho VT}.$$

Die Messung von [00,1] ist daher mit einer Unbestimmtheit verbunden von der Grössenordnung

$$\Delta[00,1]\approx \Delta p_x/\rho VT, \tag{10}$$

wo $\Delta p_x$ die Unbestimmtheit des Impulses ist. Die Dauer der Impulsmessung sei $\Delta t$ (selbstverständlich ist $\Delta t \ll T$); $\Delta x$ sei die mit der Impulsmessung verbundene Unbestimmtheit der Koordinate. Die Unbestimmtheit $\Delta p_x$ besteht aus zwei Gliedern: aus dem gewöhnlichen quantenmechanischen Glied $h/\Delta x$ und aus einem Glied, das mit dem Gravitationsfeld verknüpft ist, das von dem Probekörper selbst wegen seines Rückstosses bei der Messung erzeugt wird. Wegen der Einsteinschen Gravitationsgleichung $\Box h_{01}=\rho v_x$ muss die Unbestimmtheit von $h_{01}$, die infolge der unbestimmten Rückstossgeschwindigkeit $\Delta x/\Delta t$ entsteht, von der Grössenordnung $\rho\frac{\Delta x}{\Delta t}\cdot\Delta t^2$ sein. Es ist aus (9) ersichtlich, dass die entsprechende Unbestimmtheit von [00,1] von der Grössenordnung $\rho\Delta x$ ist, und damit während jeder Impulsmessung eine mit dem Gravitationsfeld verbundene zusätzliche Impulsunbestimmtheit von der Grössenordnung $\rho\Delta x\cdot\rho V\Delta t$ entsteht. Um den Vergleich mit den gewöhnlichen Messeinheiten zu erleichtern, sehen wir hier (bis zum Ende dieses Paragraphs) von unserer Vorschrift $c=1$, $G=1/16\pi$ ab. Wir erhalten für den Impuls

$$\Delta p_x \approx \frac{h}{\Delta x}+G\rho^2 V\Delta x\Delta t.$$

Es kann gezeigt werden (analog zu den Überlegungen von Bohr und Rosenfeld), dass das zweite Glied gegenüber dem ersten beliebig klein gemacht werden kann. Für die beste Ausführung der [00,1]-Messung scheint es aber zweckmässiger, $\Delta p_x$ zum Minimum zu machen, d.h. für die beiden Glieder die gleiche Grössenordnung zu schaffen. Dazu soll $\Delta x$ von der Grössenordnung

$$\Delta x \approx \frac{1}{\rho}\left(\frac{h}{GV\Delta t}\right)^{1/3}$$

gewählt werden. Für $\Delta[00,1]$ erhalten wir

$$\Delta[00,1] \gtrsim \frac{h^{1/2}G^{1/2}\Delta t^{1/2}}{V^{1/2}T}. \tag{11}$$

Eine absolut genaue Messung des Schwerefeldes wäre also dann möglich, wenn eine beliebig schnelle Impulsmessung möglich wäre. Zwei Umstände machen es aber unmöglich, eine beliebig schnelle Impulsmessung auszuführen: erstens soll nach der Definition der Messung $\Delta x \ll V^{1/3}$ sein, und das führt zu

$$\Delta t \gg \frac{h}{\rho^2 G V^{1/3}}.$$

Zweitens kann nach der Relativitätstheorie niemals $\Delta x$ grösser als $c\Delta t$ werden; und das führt zu

$$\Delta t \gtrsim \frac{h^{1/4}}{c^{3/4}\rho^{3/4}V^{1/4}G^{1/4}}.$$

Es folgt nach (11), dass niemals $\Delta[00,1]$ kleiner als

$$\frac{h}{\rho T V^{1/3}} \quad \text{oder} \quad \frac{h^{2/3}G^{1/3}}{c^{1/3}\rho^{1/3}V^{2/3}T}$$

gemacht werden kann. Von diesen beiden Grenzen ist für den Fall eines leichten Probekörpers ($\rho V \lesssim h^{1/2}c^{1/2}/G^{1/2}$, d. h. kleiner als etwa 0,01 mgr) die erste die einzige wesentliche. Für einen schwereren Probekörper ist die zweite die wesentlichste. Es ist klar, dass für eine möglichst genaue $[00,1]$-Messung gerade schwere Probekörper zu empfehlen sind, und das bedeutet, dass nur die zweite Grenze theoretisch wichtig ist. Wir haben schliesslich

$$\Delta[00,1] \gtrsim \frac{h^{2/3}G^{1/3}}{c^{1/3}\rho^{1/3}V^{2/3}T}. \tag{12}$$

Es ist also klar, dass im Gebiet, wo alle $h_{\mu\nu}$ klein gegen 1 sind (das ist gerade die Bedeutung des Wortes „schwach" im Titel dieser Arbeit), die Genauigkeit der Schweremessungen beliebig hoch gesteigert werden kann: denn in diesem Erscheinungsgebiet gelten die angenäherten linearisierten Gleichungen (1), es gilt also auch das Superpositionsprinzip, und es ist daher immer möglich, einen Probekörper mit beliebig hohem $\rho$ zu beschaffen. Daraus schliessen wir, dass es möglich ist, wie es z. B. diese Arbeit zu tun versucht, im Rahmen der speziellen Relativitätstheorie (d. h. wenn das raumzeitliche Kontinuum ein „Euklidisches" ist) eine durchaus konsequente Quantentheorie der Gravitation aufzubauen. Im Gebiet der allgemeinen Relativitätstheorie, wo die Abweichungen von der „Euklidizität" beliebig gross sein können, steht aber die Sache ganz anders. Denn der Gravitationsradius des zur Messung dienenden Probekörpers ($G\rho V/c^2$) soll keineswegs grösser als seine linearen Abmessungen ($V^{1/3}$) sein; daraus entsteht eine obere Grenze für seine Dichte ($\rho \lesssim c^2/GV^{2/3}$). Die Messungsmöglichkeiten sind also in diesem Gebiet noch mehr beschränkt, als es sich aus den quantenmechanischen V.-R. schliessen lässt. Ohne eine tiefgehende Umarbeitung der klassischen Begriffe scheint es daher wohl kaum möglich, die Quantentheorie der Gravitation auch auf dieses Gebiet auszudehnen.

## § 5. Wechselwirkung mit der Materie.

Ein korrespondenzmässig richtiger Ansatz für die Energie der Wechselwirkung zwischen Gravitationsfeld und Materie kann aus der von V. Fock[1] aufgestellten allgemeinrelativistischen Form der Diracschen Wellengleichung gewonnen werden. Wenn alle $h_{\mu\nu}$ klein gegen 1 sind, kann man diese Gleichung für den Fall des verschwindenden elektromagnetischen Feldes in der folgenden Form schreiben:

$$\frac{h}{i}\sum_{k=0}^{3} e_k \alpha_k \left(\frac{\partial}{\partial x_k} - \frac{1}{2}\sum_{l=0}^{3} e_l h_{kl}\frac{\partial}{\partial x_l}\right)\psi +$$

$$+\left(\frac{1}{8}\frac{h}{i}\sum_{k=0}^{3}\frac{\partial h_{00}}{\partial x_k}\alpha_k - m\beta\right)\psi = 0,$$

[1] V. Fock, ZS. f. Phys. 57, 261, 1929.

wo

$$e_0 = 1, \quad e_1 = e_2 = e_3 = -1$$

und

$$\alpha_0 = 1, \quad \alpha_1 = \rho_1\sigma_1, \quad \alpha_2 = \rho_1\sigma_2, \quad \alpha_3 = \rho_1\sigma_3, \quad \beta = \rho_3$$

(Diracsche Matrizen). Führen wir statt der vierkomponentigen $\psi$-Funktion zwei zweikomponentige Funktionen $\chi$ und $\varphi$ ein

$$\psi_1 = \chi_1 e^{-imt/h}, \quad \psi_2 = \chi_2 e^{-imt/h}, \quad \psi_3 = \varphi_1 e^{-imt/h}, \quad \psi_4 = \varphi_2 e^{-imt/h},$$

so wird bei abnehmender Geschwindigkeit des Teilchens $\chi$ gleich Null, und $\varphi$ zu seiner nichtrelativistischen Wellenfunktion. Es lässt sich aus der Schrödingergleichung für diese $\varphi_x$ ersehen, dass die Energie der Wechselwirkung zwischen dem Teilchen und dem Gravitationsfeld die Form

$$V = \frac{m}{2} h_{00} + \frac{h}{i} \sum_k h_{0k} \frac{\partial}{\partial x_k} - \frac{h^2}{2m} \sum_{kl} h_{kl} \frac{\partial^2}{\partial x_k \partial x_l} - \frac{h^2}{4m} \sum_{kl} \frac{\partial h_{kl}}{\partial x_l} \frac{\partial}{\partial x_k} +$$
$$+ \frac{h}{4i} \sum_l \frac{\partial h_{0l}}{\partial x_l} + \frac{h^2}{4mi} \sum_{jklm} \sigma_j e_{jlm} \frac{\partial h_{km}}{\partial x_l} \frac{\partial}{\partial x_k} + \frac{h}{4} \sum_{jlm} \sigma_j e_{jlm} \frac{\partial h_{0m}}{\partial x_l}$$

hat, wo $e_{jlm}$ der schiefsymmetrische Einheitstensor ist (d. h. $e_{123} = 1$ und $e_{jlm}$ ist in Bezug auf jedes Paar seiner Indizes antisymmetrisch). Wenn die Wellenlänge der Gravitationsstörungen genügend gross ist, vereinfacht sich dieser Ausdruck zu

$$V = \frac{m}{2} h_{00} + \frac{h}{i} \sum_k h_{0k} \frac{\partial}{\partial x_k} - \frac{h^2}{2m} \sum_{kl} h_{kl} \frac{\partial^2}{\partial x_k \partial x_l}. \tag{13}$$

Den Ansatz (13) werden wir im folgenden benutzen. Es sei bemerkt, dass sogar die einfachen korrespondenzmässigen Betrachtungen, ohne den Umweg über die Dirac-Focksche Gleichung, auch zum Ansatz (13) für die Wechselwirkungsenergie führen.

## § 6. Energieübertragung durch die Gravitationswellen.

Eine der einfachsten Anwendungen der oben skizzierten Quantentheorie der Gravitation besteht in der Berechnung der Energieausstrahlung in Form der von materiellen Syste-

gurationsraum zurückgehen, erhalten wir demgemäss eine Schrödingergleichung mit der potentiellen Energie

$$-\frac{m_1 m_2}{16\pi\,|\mathfrak{r}_1 - \mathfrak{r}_2|},$$

und wir haben also das Newtonsche Gravitationsgesetz als eine notwendige Folgerung der Quantentheorie der Gravitation wiedergefunden.

Physikalisch-technisches Institut
und Physikalisches Institut der Universität.
Leningrad, August 1935.

# Appendix 2.
# M.P.Bronstein's note

"Über den spontanen Zerfall der Photonen", 1936

# Briefe, vorläufige Mitteilungen und Diskussionen.

## ÜBER DEN SPONTANEN ZERFALL DER PHOTONEN.

*Von M. Bronstein.*

(Eingegangen am 2. November 1936.)

In einem bekannten Aufsatz, wo zuerst über die Möglichkeit der Streuung von Licht an Licht die Rede war, hat Halbern[1] auch zwei andere Hypothesen geäussert, nämlich dass sich (1.) infolge der Wechselwirkung zwischen dem Licht und den auf den negativen Energieniveaus befindlchen Elektronen jedes im leeren Raume sich bewegende Photon in zwei oder mehrere Photonen von gleicher Gesamtenergie und gleicher Richtung spontan verwandeln kann, und (2.) dass vielleicht die von Hubble entdeckte kosmiche Rotverschiebung gerade durch einen solchen spontanen Photonenzerfall erklärt werden kann (das allmähliche Absplittern kleiner ultraroter Photonen von einem durch den interstellaren Raum reisenden Photon). Die erste Hypothese von Halpern vertritt auch Heitler[2] in seinem unlängst erschienenen Buch; ausserdem behauptet er, ohne es zu beweisen, dass ein Photon spontan nur in drei und nicht z. B. in zwei Teile zerfallen kann.

Man kann, ohne den speziellen Mechanismus des Photonenzerfalls in Betracht zu ziehen, leicht zeigen, dass die zweite Halpernsche Hypothese sicher falsch ist. Wenn in einem Bezugssystem ein Photon in irgendeinem Verhältnis sich spaltet, dann spaltet es sich von jedem anderen Bezugssystem aus betrachtet auch in demselben Verhältnis, d. h. in die gleiche Zahl Teile mit derselben relativen Energieverteilung; denn infolge des Dopplerprinzips multiplizieren sich alle Frequenzen beim Übergang zum zweiten Bezugssystem mit ein und demselben Faktor. Wenn die Wahrscheinlich-

---

[1] O. Halpern, Phys. Rev. **44**, 885, 1934.

[2] W. Heitler, The Quantum Theory of Radiation. Oxford 1936, S. 193.

keit des spontanen Zerfalls eines bestimmten Photons in einem gegebenen Verhältnis pro Zeiteinheit $w$ ist, dann ist $w(t_B - t_A)$ relativistisch invariant, wobei $t_B - t_A$ die Zeitspanne zwischen zwei Punkten $A$ und $B$ der Photonenweltlinie bedeutet. Nun ist gegenüber den Lorentztransformationen auch $(t_B - t_A)/\nu$ relativistisch invariant, wo $\nu$ die Frequenz ist. Es folgt daraus, dass $w\nu$ eine Konstante oder mindestens eine Funktion von $\Delta\nu_x \Delta\nu_y \Delta\nu_z/\nu$ ist. $\Delta\nu_x$, $\Delta\nu_y$, $\Delta\nu_z$ sind die Unbestimmtheiten der Komponenten des Wellenvektors des Photons; $\Delta\nu_x\Delta\nu_y\Delta\nu_z/\nu$ ist die einzige mit dem Photon verbundene Lorentzinvariante. Nun zeigt die Beobachtung, dass die relative Rotverschiebung $\Delta\lambda/\lambda$ für alle Wellelängen im Spektrum eines und desselben Spiralnebels genau gleich ist, anderenfalls hätten die Astronomen z. B. aus dem roten Ende des Spektrums eine und aus dem violetten Ende eine andere Geschwindigkeit eines und desselben Spiralnebels berechnet. Diese triviale Tatsache würde uns zu der Folgerung zwingen, dass, wenn die Halpernsche Hypothese richtig wäre, $w$ und nicht z. B. $w\nu$ von der Frequenz $\nu$ unabhängig sein sollte, was mit den oben genannten Eigenschaften der Lorentztransformationen im krassen Widerspruch steht.

Was die erste Halpernsche Hypothese betrifft, so zeigen die Rechnungen, dass die Wahrscheinlichkeit des spontanen Photonenzerfalls von Null verschieden wird, wenn wir von der Polarisation des Vakuums absehen und in der Wechselwirkungsenergie zwischen Photonenfeld und Elektronenfeld nur das erste Glied, das der ersten Potenz der Elektronenladung proportional ist, berücksichtigen. Es zeigt sich in diesem Fall, dass die spontane Zersplitterung im Einklang mit der Heitlerschen Behauptung nur in drei Teile möglich ist, wobei für die Berechnung die vierte Näherung der Störungstheorie nötig ist. Wenn wir aber die Polarisation des Vakuums in Betracht ziehen, d. h. in dem Wechselwirkungsenergieausdruck auch die von Heisenberg berechneten „Subtraktionsglieder" abziehen, dann zeigt es sich, dass die Wahrscheinlichkeit des spontanen Zerfalls streng gleich Null wird. Vom physikalischen Standpunkte aus ist es natürlich und befriedigend, denn andernfalls würde diese Wahrscheinlichkeit mit abnehmender Frequenz zunehmen, was mit dem Korrespondenzprinzip und der Erfahrung im Widerspruch wäre.

Wir kommen also, im Gegensatz zu Halpern und Heitler zu dem Schluss, dass der spontane Photonenzerfall überhaupt unmöglich sei.

Die dazu gehörigen Rechnungen werden an einer anderen Stelle veröffentlicht werden.[1]

Leningrad.
Oktober 1936.

[1] ZS. f. exp. und theor. Phys., russisch, noch nicht ersch.

# Appendix 3.

M.P. Bronstein

## Inventors of Radiotelegraph

(First chapters of the book)

### Who and When?

Who invented the radio and when did it happen?

Some people would say that it was Alexander Popov who invented the radio forty years ago [the book was written in 1936 – *Ed.*], others believe that it was Guglielmo Marconi.

Indeed, some forty years ago both Popov and Marconi simultaneously created the first ever radio stations and sent out the first ever radiotelegrams.

Yet the history of the radio goes further back than the first radiotelegrams. The scientists whose discoveries and experiments launched it never transmitted radiotelegrams. What was more they never tried to transmit either signals or human speech. They would have been amazed had they been told that they were inventing radiotelegraph!

They were not concerned with transmitting sounds, signals or images. They were engrossed in different things.

Have you ever seen electric charges sparkling in electrified things? Have you seen how they flashed to extinguish the next moment? The history of the radio started with these sparkles.

For many decades physicists had observed electric sparks, experimented with them and studied their properties. At last they asked themselves: how long is the interval between the spark's birth and death? How long does it live? This was not an easy question: the spark flared up and went out at once. How long exactly was this interval? A hundredth fraction of a second or even less – a thousandth or a millionth? How could one measure it?

Everyone feels the flow of time. All of us are able to distinguish between one and two minutes, one and two seconds or even the second's tenth fraction and one second. Man cannot tell apart anything less than one-tenth or one-fifteenth fraction of a second – they seem to be the same as a hundredth, thousandth and millionth fraction. Our senses are not precise, swift and subtle enough. We fail to perceive a time interval that is smaller than one-fifteenth. That is why a hundredth and a thousandth or a thousandth and a millionth fraction seem equal to us. We call them a moment.

And what about the clock? After all, clocks and watches were invented to measure time. Can the watch measure a moment?

Let us look into a precision instruments plant. There are all sorts of clocks and watches there including chronometers that sailors take on long sea voyages, super-exact astronomical watches, electrical chronographs and stopwatches. Yet there are no watches for measuring the second's millionth fractions.

But this watch does exist – it was invented 75 years ago by a German physicist Wilhelm Feddersen to measure the electric spark's life-span.

Little did he think that he was launching the history of the radio.

## Feddersen's Clock

Today his invention can be seen in a museum in the German city of Munich.

It does not look like an ordinary clock – it has no hour or minute or second hand. Indeed, what hands are needed to measure the second's millionth parts? What sort of a hand is able to move so that an eye can register at a speed of a million steps per second? Feddersen's task was to make these steps apparent.

It took him long time to find a suitable hand for his watch. He made it out of the material that had not been used for clocks and watches before him – it was neither steel nor bronze.

He made it out of light rays.

Take a pocket mirror and bring it out into the sunshine. It will reflect the sun rays as a bright light spot.

As we move the mirror, the spot will swiftly dance and jump. Here it is close and the next moment it is on the other side of the street dancing on the walls and balconies.

The light spot moves quickly – a hundred or even a thousand times faster than the clock's second hand. Can we force it to move a million times faster?

Yes, we can. We should spin the mirror faster. It is better to use a machine to rotate the mirror – it is not only swifter than the human hand but it is also more exact: you can order the speed with which you want the mirror to spin.

In Feddersen's machine there was a heavy weight that pulled the rope as it was lowering. The rope rotated the shaft with a gear haft onto it. The first gear moved the second gear; the second gear set the third one into motion while the third agitated a large steel screw. As the motion progressed from the first gear to the screw it became faster and faster: the first gear was not very fast, it made several turns per a second; the second gear was faster; and the third faster still. The steel screw was the fastest: it spun one hundred times per second.

To prevent the device from shaking as it was gaining speed, Feddersen fixed it to a firm foundation. He placed two solid cast iron beams in a wall and attached a heavy iron box to them which was open from the front and the sides. The screw pierced the box's bottom and lid.

Then he had to supply the screw with a mirror that would reflect the light rays. To make it Feddersen bought two concave lenses that are normally used make glasses for near-sighted people. He covered them with silver on one side and got two small concave mirrors. He then fixed them to the screw so that they looked in opposite directions. As the weight was going down, the screw moved and the mirrors were rotating steadily and swiftly.

This was the mechanism of the clock Feddersen invented. It was not enough though. To make the clock complete one needs also a face on which the path travelled by the hand is measured. If the hand is made of light what should the face be made of?

It was a task that took much time to solve. Feddersen used a light-sensitive photoplate to serve the face of the unusual clock he had invented. The electrical spark would mark the beginning and end of its short life with its own rays. They would be reflected in the rotating mirror, the swift light spot would travel across the photoplate and leave its trace on it. The longer the spark would burn the longer the spot's trace. It is easy to calculate the time the light spot travelled across the plate if we know its speed – this will be the electrical spark's life-span.

The gadget was ready. The steel screw was fixed in the cast-iron frame, the mirror spun regularly, the photoplate was put in place. The Leyden jar that would produce sparks was also prepared. It consisted of three jars fitted one into the other: the inside and outside jars were made of metal with the middle jar made of glass.

Feddersen charged the jar: the outside metal jar was charged with positive electricity and the inside metal jar with negative. Then he connected the jar with two metal balls placed opposite one another; they were called the spark gap. Feddersen connected the inner jar with one ball and the outside jar with another. What was left was to press the button that would complete or break the circuit. The electrical charge will travel along the wires. The positive charge will meet the negative charge and a bright sparkle will appear between the balls.

## The Story of a Spark

Feddersen started experimenting with his unusual clock. He lowered the weight that set the gears and the screw in motion. The screw and the mirror began to spin so quickly that it glittered before him. The fly-wheel attached to the screw whined at a high pitch as it was rotating. Feddersen listened – the pitch was always the same which meant that motion was steady with no acceleration or deceleration.

Then he extinguished the light and opened the cassette with the plate; he pushed the button that completed the electric circuit. Immediately there was a spark between the balls: the electrical charge travelled from one metal jar to another. In the twinkling of an eye it travelled along the wires and sparkled as it made its way from one ball to the other.

It lasted a very short moment, yet the swift reflection in the mirror managed to reach the photoplate. It travelled across the plate with the velocity of an artillery shell and left its trace.

Feddersen immediately developed the plate and printed a photo. It clearly showed a narrow stripe – the trace of the light spot.

The physicist measured its length – it was one and a half centimeters; the light spot's velocity was 60,000 cm per second. How much did it take the light spot to travel across the plate? To find this out one should divide the length of the trace (1.5 cm) by the spot's velocity (60,000 cm/sec). The result was 0,000025 sec.

This means that the spot was travelling across the plate for 25 millionth fractions of a second, and this was the electrical spark's life span.

So the Feddersen clock stood the test: it measured the life-span of a spark; Feddersen solved the problem he had put to himself.

While closely examining the photo Feddersen realized that the clock had made another discovery. It had not only measured the spark's life but it also had shown *what* it was. The photo was, in fact, the spark's detailed biography.

The trace of the light spot proved to be broken rather than continuous. It consisted of several lighter places divided by dark spaces.

This means that the spark which appeared between the two balls was not shining continuously. It flashed and extinguished several times during its short life of twenty-five millionths of a second. The flashes followed one another with the speed that the human eye could not distinguish between them and took them for one spark. It was only Feddersen's miraculous clock that managed to divide them into several flashes.

Feddersen counted the flashes in his photo: there were eight of them and each was a little bit brighter than the following one. So, there were eight flashes in twenty-five millionths of a second. Seemingly the spark consisted of separate flashes that appeared and were extinguished in three millionths of a second!

This was how Feddersen managed to read the life story of a spark recorded by a light spot that lasted one moment.

## Experiments Go On

Feddersen repeated his experiment many times. He would take one Leyden jar or many of them (10, 15 and even 20). He would move the balls close to each other or apart with a centimeter or a centimeter and a half between them. He tried iron, copper, lead and gold balls; he used the balls made of the same and of different metals. He experimented with the wires that connected the jars with the balls – they were short and thick or long and thin. Every time he photographed the spark's reflection in the spinning mirror.

From the photographs he learned how long the spark lasted, how it flared up, burned and was extinguished.

He got sparks of different lengths, different life-spans and different brightness but all consisted of more than one flash. They followed one another with the intervals of several millionths of a second, grew weaker until they disappeared altogether.

The question was: why did the electrical charge made several leaps between the balls rather than jumping between them only once?

While turning over all possible answers in his head, Feddersen recalled an article about the discharges in the Leyden jar. It was written in 1853 by William Thomson, English scientist. He had never experimented with the Leyden jar; what was more he had never experimented at all. He was a skillful mathematician who knew the physical laws that guided electrical current and was able to make mathematical deductions from them. He set out to establish, through calculations, what happened to the electrical charge when the Leyden jar was discharged.

Thomson's calculations demonstrated that the electrical charge, having travelled from the jar to the balls along the wire jumped across the leap and travelled along the wire back to the jar. This time it went to another jar. Obviously, the positive and negative charges changed places.

The negative charge that at first resided in the inside jar would remove itself to the outside jar, while the positive charge would travel in the opposite direction. Thus, the jar would never be discharged – it was charged in a new way. The electrical current would travel to the balls in the opposite direction. The electrical charge would travel from one ball to the other.

"That was what happened!" Feddersen thought. That was why the clock registered not one but eight sparks. From this it followed that the charge jumped between the balls from one side to the other and back eight times. Every time there was a bright flash. As soon as one flash died another appeared – the electrical current was renewed and travelled in the opposite direction. Flash after flash sparkled in the narrow gap between the balls while the spark was alive and the current changed its direction with every new flash. There was a flare-up when it traveled in one direction and another flare up when it traveled in the opposite direction.

Feddersen's spinning mirror proved that Thomson was right: the electrical spark was a short section of alternating current. It changed its direction in negligible spans of time, in several millionths fractions of a second.

This was how the spinning mirror helped Feddersen study the nature of the electric spark ...

## About Matvei Bronstein's Last Book

Just like his earlier children's books this one was also first published in the magazine *Koster* (Bonfire). Just like his earlier stories about helium and *X*-rays it was to be published under a separate cover. The author and the editor carefully checked and rechecked every word, the designer supplied it with illustrations, the printers did what they should. The book was really ready – it was waiting to be bound but ... In spring 1937 the entire print run of tens of thousands of copies was destroyed.

What happened? Why was this book found wanting?

To answer this question one has to consider a seemingly different topic: how science is advanced.

Matvei Bronstein began his last book with a question: who invented the radio and when did it happen? Those not familiar with the way science is developing might imagine that a certain name and a certain date would suffice. There is no straightforward answer in his book – in fact, the whole book is an answer to this question. Even this long answer is not enough. Feddersen's remarkable light clock was not invented out of nothing: it was thirty years earlier that another physicist had suggested to use a spinning mirror as a clock. There are many such examples.

Great Newton used to say: "If I saw farther than the others it was because I had the giant's shoulder to mount on". This can be applied to all "who see farther than the others".

Anybody wishing to make a step forward in science has to travel along the road that took the pioneers a life time. The concepts, experiments and merely designations, gained with great efforts by scientists of the past, shorten the way for those who came after them. Science requires a collective effort; without it science would have been impossible.

This is not obvious to those who have never traveled along these roads yet are bold enough to pass judgment on those who have. Granted freedom of action such people can do much harm behind a smoke screen of high-sounding words about truth and patriotism. In fascist Germany they were digging for the purely Aryan roots in every science. In Soviet Russia there was a similar period when the bosses of all ranks demanded Russian roots in all branches of science and technology – from a bicycle to the law of conservation.

It was in 1937 when precisely such a person was appointed director of the publishing house where Bronstein's book was being prepared for press. The boss disliked the book at first glance: could it be true that the famous Russian inventor had used what scientists from other countries discovered before him? He could not fathom how two scientists living in different countries could simultaneously and independently make one and the same discovery! Obviously, one of them was a thief and it took no wisdom to guess who.

The boss invited the author into his office and demanded that "corrections" be introduced into the text. In vain did Bronstein try to explain that these things were nothing out of the ordinary in the history of science: after all, the physicists were studying the same nature and reading the same articles. He cited examples: Janseen and Lockyer who independently discovered the yellow line in the helium specter. Why couldn't Popov and Marconi discover the radio, he argued.

It seems that the boss was unimpressed: he wanted to read in the book that it was a Russian inventor alone who discovered the radio. He even went farther and said: a true Soviet patriot should defend Russian priority everywhere and at any price.

When Bronstein realized that the man was totally indifferent to the roads science treaded and insisted that he would lie "out of patriotism"; he said that it was fascist patriotism and that he had no desire to forge history. With these words he left the office.

The book was sent to a paper shredder that reduced it to a heap of paper strips. Several months later the author was executed in the basement of a Leningrad prison.

# Photographs

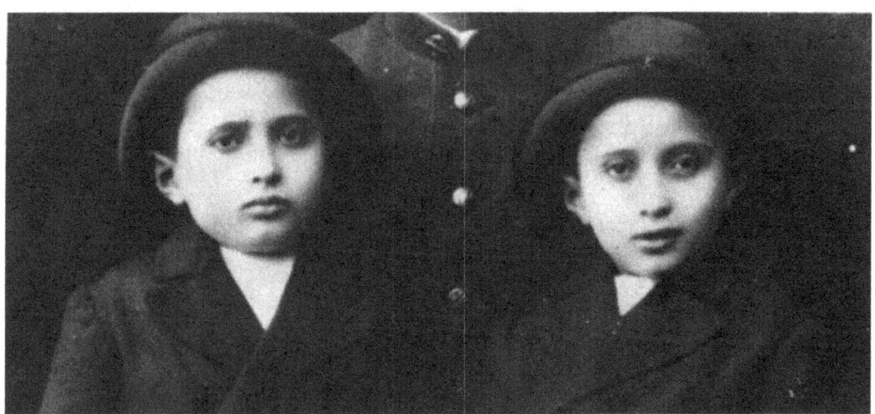

Matvei (left) and Isidor at six

"Here is what I look like now: naturalism in the highest degree, up to unshaven cheeks (photo for a travel card). 1928". (Inscription on the reverse side)

Matvei, Mikhalina and Isidor Bronstein. Kiev, summer of 1928

V. Ambartsumyan, N. Kozyrev, M. Bronstein and I. Kibel before their Armenian journey. Summer of 1929

Matvei with his parents, sister and nephew

In the sisters Kanegissers' flat (from left to right): L. Landau, E. Kanegisser, V. Ambartsumyan, unidentified person, N. Kanegisser, M. Bronstein

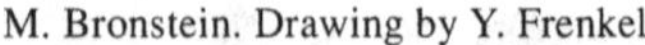

M. Bronstein. Drawing by Y. Frenkel

In 1931 there was the First All-Union Conference on Planning in Science. The drawing shows how Bronstein treated the subject. (Inscription on the drawing: "to plan is to tell fortunes".)

M. Bronstein delivers a lecture on the theory of gravitation

M. Bronstein delivers a lecture on quantum mechanics

A. Ioffe

P.A.M. Dirac

G. Gamow

F. Joliot

Humorous drawings by N. Mamontov of participants at the First Nuclear Conference, Leningrad, September 1933

V. Fok

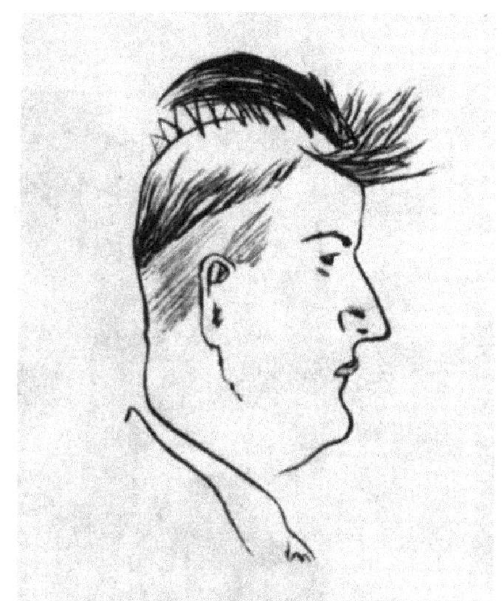

I. Kurchatov

Y. Frenkel

M. Bronstein

One of the last photos of M. Bronstein

# Index

# Bibliography after 1994


1. *Gorelik G.*, 'Meine antisowjetische Tätigkeit …'. Russische Physiker unter Stalin. (Transl. Helmut Rotter). Vieweg, 1995.
2. *Gorelik G.*, Lev Landau, prosocialist prisoner of the Soviet state // Physics Today, 1995, May, p. 11–15.
3. *Gorelik G.*, The Top Secret life of Lev Landau // Scientific American, 1997, August, p. 72–77.
4. *Gorelik G. with A. Bouis*, The World of Andrei Sakharov. A Russian Physicist's Path to Freedom. Oxford University Press, 2005.
5. *Gorelik G.*, Matvei Bronstein and quantum gravity: 70th anniversary of the unsolved problem // Physics-Uspekhi 2005, vol. 48, no. 10, pp. 1039–1053.
6. *Gorelik G.*, Sovetskaya zhizn Lva Landau. [The Soviet Life of Lev Landau]. Moscow: Vagrius, 2008.
7. *Gorelik G., Shalnikova N.* (eds)., Sovetskaya zhizn Lva Landau glazami ochevidtsev. [The Soviet Life of Lev Landau in eyewitness accounts] Moscow: Vagrius, 2009.
8. *Gorelik G.*, The Paternity of the H-Bombs: Soviet-American Perspectives // Physics in Perspective, Vol. 11, N. 2 / June, 2009, p. 169–197.
9. *Gorelik G.*, Andrei Sakharov: Nauka i svoboda [Andrei Sakharov: Science and Freedom]. Revised 3rd ed. Moscow, Molodaya gvardiya, 2010.
10. Web presentation. Matvei BRONSTEIN (1906–1938) http://people.bu.edu/gorelik/MPBronstein_100/MPBronstein_100.htm